I Leslie Rubin
Joav Merrick
Editors

Environmental Health

Leaders for Climate Change

nova science publishers
www.novapublishers.com

DOI: https://doi.org/10.52305/TBTI6175

Library of Congress Cataloging-in-Publication Data

ISBN: 979-8-89530-347-4 (Hardcover)
ISBN: 979-8-89530-393-1 (eBook)

Published by Nova Science Publishers, Inc. † New York

Contents

Chapter 1

Break the Cycle of Climate Change

I Leslie Rubin[1-5,*], MD
Abby Mutic[3,6], APRN, PhD
Victoria Green[3,7], JD, MD
Melissa Gittinger[3,8], DO
Jinbing Bai[3,6], PhD, RN
Nathan Mutic[3,6]
Wayne Garfinkel[3,9], BSCE
Henry Falk[3,4], MD, MPH
Benjamin A Gitterman[10,11], MD
and Joav Merrick[12-15], MD, MMedSci, DMSc

[1]Department of Pediatrics, Morehouse School of Medicine, Atlanta, Georgia, United States of America
[2]Department of Pediatrics, Emory University School of Medicine, Atlanta, Georgia, United States of America
[3]Southeast Pediatric Environmental Health Specialty Unit, Emory University, Atlanta, Georgia, United States of America
[4]Break the Cycle of Health Disparities, Inc. Atlanta, Georgia, United States of America
[5]The Rubin Center for Autism and Developmental Pediatrics, Atlanta, Georgia, United States of America
[6]Nell Hodgson Woodruff School of Nursing, Emory University, Atlanta, Georgia, United States of America

* ***Correspondence:*** I Leslie Rubin, MD, Associate Professor, Department of Pediatrics, Morehouse School of Medicine, Director, Break the Cycle Program of the Southeast Pediatric Environmental Health Specialty Unit at Emory University, Founder, Break the Cycle of Health Disparities Inc, Medical Director, The Rubin Center for Autism and Developmental Pediatrics. 750 Hammond Drive, Building 1, Suite 100, Atlanta, GA 30342, United States. Email: lrubi01@emory.edu.

In: Environmental Health: Leaders for Climate Change
Editors: I Leslie Rubin and Joav Merrick
ISBN: 979-8-89530-347-4

[7]Department of Gynecology and Obstetrics, Emory University School of Medicine, Atlanta, Georgia, United States of America
[8]Georgia Poison Center, Grady Health System, Atlanta, Georgia, United States of America
[9]Retired EPA Region 4 Children's Environmental Health Coordinator, Atlanta, Georgia, United States of America
[10]Departments of Pediatrics and Public Health, George Washington University Schools of Medicine and Public Health and Health Services, United States of America
[11]General and Community Pediatrics, Children's National Medical Center, Washington DC, United States of America
[12]National Institute of Child Health and Human Development, Jerusalem, Israel
[13]Division of Pediatrics, Hadassah Hebrew University Medical Center, Mt Scopus Campus, Jerusalem, Israel
[14]Kentucky Children's Hospital, University of Kentucky College of Medicine, Lexington, Kentucky, United States of America
[15]Center for Human Development, School of Public Health, Georgia State University, Atlanta, Georgia, United States of America

Abstract

The consequences of poverty and racial disparities on the health and well-being of children are compounded and amplified by the impact of climate change in the form of heat, drought, flooding, and other extreme weather events. Poor children and children of color currently suffer disproportionately from the social, economic, environmental, and political realities in which they live, and are at greater risk for morbidity and mortality. The challenges are complex and multifaceted, requiring creative and coordinated responses at multiple levels. There are two fundamental challenges of undeniably gargantuan proportion: first the ever-increasing rise in global temperature without a sign of respite that requires a concerted effort to dramatically reduce greenhouse gases from the burning of fossil fuels; and second the intractable historic social, economic, and racial inequities, locally and globally that require a concerted effort to confront the historic injustices and develop strategies to promote justice and equity for all. There is no one solution to these challenges, but it is undeniable that a global concerted effort is needed, whether it be from large governmental entities, intermediate size public and private agencies and organizations, or small groups and individuals. In this book we bring some research and perspectives on the impact on the health of our children in this world of climate change and how we can make a positive difference.

Introduction

Most historians place the origin of the Industrial Revolution in Great Britain in the middle decades of the 18th century, with changing patterns of human settlement from agriculture to industry, affecting labor, family life and the populating of cities (1). There was a surge of industrial development in the 19th century with a burst during the first half of the 20th century that involved two world wars, and the employment of the heavy industry required to engage in those deadly battles. By the mid-20th century, there was another explosion of industrial activity in response to research and technological developments, the rise of global demands for convenience, comfort, and security, and the increase in competition at political and economic levels. The finances, energy and resources that were poured into these efforts resulted in an unprecedented surge of global progress in technology, transportation, communication, and consumer products that could be made readily available to those who could afford them.

To make this surge happen, the industrial process required fuel. This fuel came from the depths of archeological history, when plant and animal carbon were compacted over millions of years into the hard rock of coal and the thick fluid of oil. The use of these fuels to power and feed the hungry maw of industry required burning, which released gases and particulate matter with many chemicals of long past life forms, most notably carbon and sulfur. These elements, when burned, combined with oxygen to form gases and were released in large quantities into the atmosphere. While the release of these potentially toxic gases and compounds into the atmosphere may have seemed trifling in the early days, the accelerated burning of fossil fuels to meet the growing needs of industry resulted in the release of increasingly large amounts of these gases and toxins into the air, the water and the soil. This avalanche of toxic gases and other by-products of industry has been linked to an increasing burden of human illness (2).

The accumulation of the by-products of human consumption into the air and atmosphere created a blanket of gases that prevented the dissipation of heat generated on the planet to the outer stratosphere, thus trapping the heat resulting in an increase in the earth's temperature. This increase has been recorded and noted to be in excess of what had been predicted and, disturbingly, continues to increase. According to the National Aeronautics and Space Administration (NASA), Earth's global average temperature has increased by about 1 degree Celsius (1.8 degrees Fahrenheit), since the pre-industrial period, and they predict that the temperature will continue to rise by more than 0.2 degrees Celsius (0.36 degrees Fahrenheit) per decade (see Figure 1). Furthermore, NASA clearly attributes the current warming trend to human industrial activity since the 1950s (3). Due to the significant trend of global warming across the globe, the impact on a long-term shift in temperatures and weather patterns is referred to as climate change. For example, the increase in temperatures and associated drought conditions increase the likelihood of wildfires that consume millions of acres of forests, thus reducing tree coverage and its benefit in reducing carbon dioxide by extraction from the atmosphere. The increase in temperature and reduced green spaces further serve to increase global warming and associated climate change, ultimately disrupting the ecology of the planet and adversely affecting human health (4).

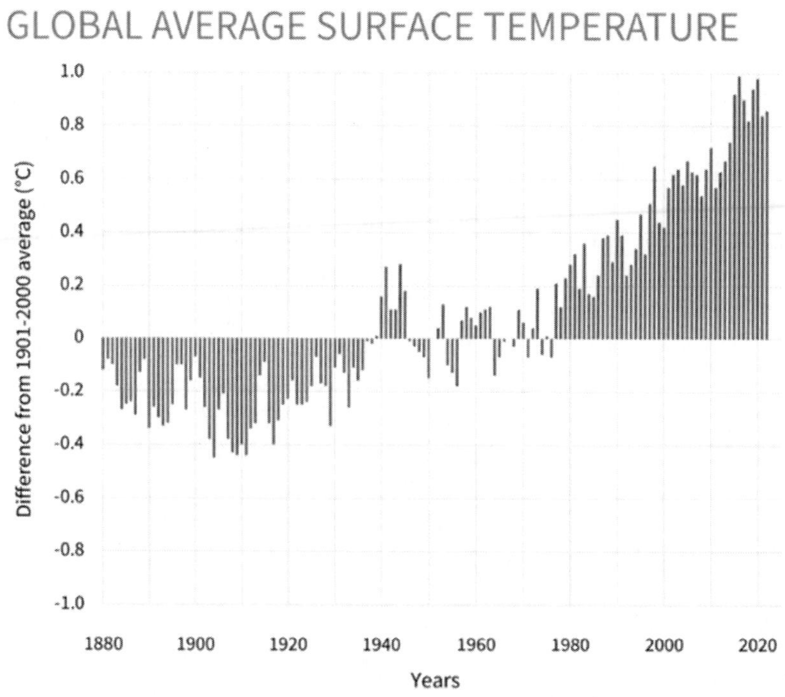

Figure 1. Yearly surface temperature compared to the 20th-century average from 1880–2022. Bars below the baseline indicate cooler-than-average years; bars above the baseline show warmer-than-average years. NOAA Climate.gov graph, based on data from the National Centers for Environmental Information (5).

Global warming is associated with an increase in ambient temperature globally. Given these changes affect ocean temperatures, polar icecaps melt at a rapid rate, resulting in the expansion of water volume and rises in sea level. The US National Oceanic and Atmospheric Association (NOAA) reports that sea level rose about four to five inches between 1900 and 1990, with an accelerated increase of 3 inches between 1990 and 2015, and a predicted rise in sea level of one-eighth of an inch per year (6). The realities of coastal flooding are already felt around the world, with catastrophic floods in Pakistan in 2022. The floods affected 33 million people, with a loss of more than 1,700 lives. Most water systems in affected areas were damaged, forcing more than 5.4 million people, including 2.5 million children, to rely solely on contaminated water from ponds and wells. Since then, children have been suffering from infectious diseases, especially gastroenteritis, which, in the pathophysiological context of their chronic malnutrition, results in an increased mortality rate. UNICEF reports that more than 6 months after the disaster, more than 20 million people, almost half of whom are children, continue to need humanitarian assistance (7).

Warmer temperatures have destabilized the jet stream, resulting in an increased frequency and intensity of extreme weather events. In their review of extreme weather events in the US during the year 2022 (see Figure 2), NOAA reported 15 natural disasters that cost more than a billion dollars each, being the 8th consecutive year that there have been 10 or more disasters costing more than a billion dollars (8). In addition to the financial costs of natural disasters, there have been many environmental costs: 7,472,995 acres of forest have been consumed by wildfires; a record nine inches of rain in one day was recorded in St. Louis in the summer of 2022; a 2.5 degree Fahrenheit increase in average US summer temperatures has been recorded; the most severe and prolonged drought in 1,200 years has begun; an Arctic blast has swept across the US in December of 2022, disrupting power supply to 1.5 million customers; and 20,000 flights have been canceled at the height of holiday travel (9).

The death toll attributed directly to climate and weather disasters was recorded as 342. According to NOAA, since 1980, the U.S. has sustained 348 weather and climate disasters with overall damages/costs at or above $1 billion with a total cumulative cost exceeding $2.510 trillion (8). The unfortunate reality is that this is only going to increase.

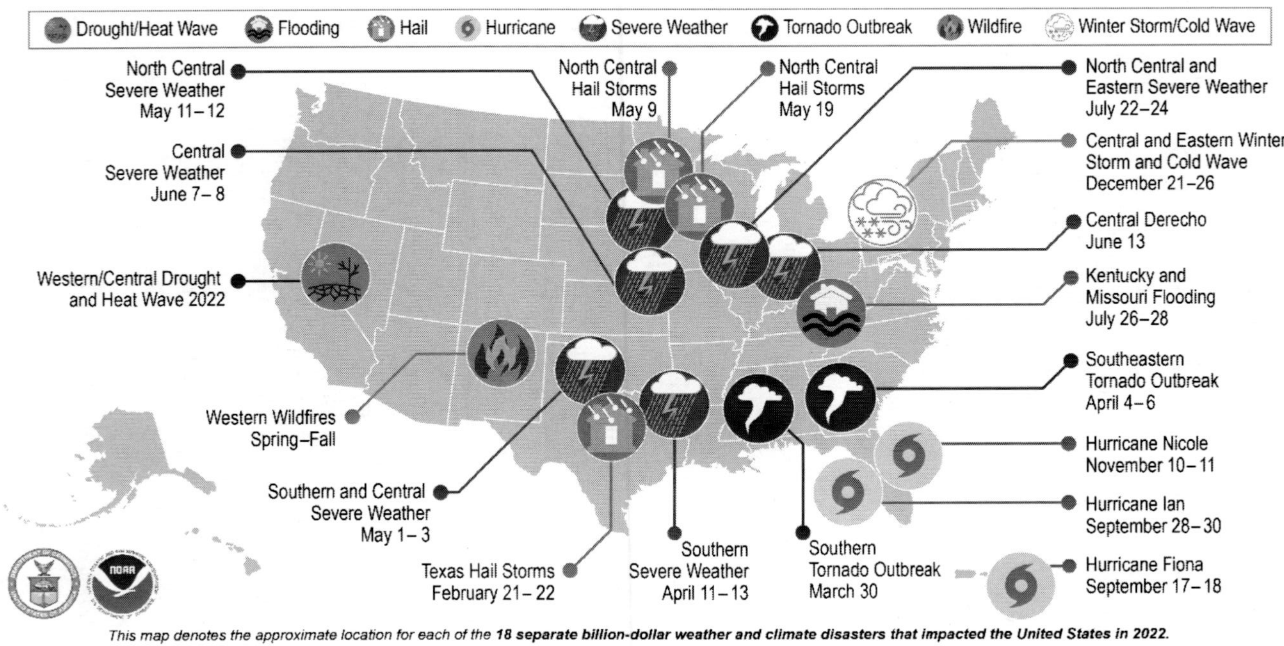

Figure 2. US 2022 billion-dollar weather and climate disasters (8).

Climate Change and Children's Health

While the impact of global warming and the associated increase in frequency and intensity of adverse weather events affect the ecology of the planet and all living beings, children are most vulnerable because of their developmental and physiological immaturity and their dependence on adults to nurture them and protect them from harm. Children are at greater risk for a variety of heat-related illnesses, such as fever and electrolyte imbalance with renal disease due to dehydration, and a significant mortality for children under 14 years of age, especially those who are most vulnerable under one year of age (10). Infectious diseases are also associated with an increase in ambient temperature, particularly gastrointestinal disorders with dehydration and electrolyte disorders, respiratory disorders such as asthma—which is related to the increase in concentration of particulate matter in the air—an increase in injuries from extreme weather events, as well as an increase in endocrine, nutritional, and metabolic disorders (11). There is also an appreciable increase in bacterial infections such as otitis media, viral infections, and disturbingly, vector borne infections such as Lyme disease, West Nile virus and malaria as the insects thrive the warmer climates. Taken together, temperatures above the minimum morbidity temperature accounted for an estimated 11.8% of Emergency Department (ED) visits for any cause and 31.0% of ED visits for heat-related illnesses during the warm season (12).

Not only do the environmental impacts of climate change affect children physically and physiologically but these impacts can also affect them psychologically and emotionally. An increase in ambient temperature has been estimated to affect children's concentration and learning, which is statistically projected to result in a 4% to 7% reduction in annual academic achievement per child, which in turn, can affect a child's future employment opportunities and projected income (13). Furthermore, children who experience climate disasters with destruction, displacement or loss associated with the disasters may experience symptoms of post-traumatic stress disorder (PTSD) with anxiety, depression, and other emotional and behavioral consequences (14-16). For many children, awareness of the existential threat of climate change, with the apparent inaction by leaders in authority to address it in a meaningful, substantial, and reassuring manner, induces anxiety and fearas well as a sense of helplessness that can occupy their thoughts and affect their ability to communicate, interact, socialize and learn (13). A projection cited by the US Environmental Protection Agency (EPA) report on children's health and the environment is that, if no additional adaptation actions are taken, 1 to 2 million children, or more, are anticipated to experience temporary home displacement or complete home loss, from coastal flooding at 50cm to 100cm of global mean rise in sea level (13).

Climate Change and Child Health Disparities

What this data does not show is the people most affected by global warming, climate change and associated extreme weather events. Obviously, those geographically in the direct path of a storm or those directly impacted by each of the extreme weather phenomena described above are affected, but it is the more vulnerable members of the community that will inevitably suffer the most: the children, elderly, and those with disabilities or chronic medical conditions. Moreover, at the population and community level, the most vulnerable communities are those who are poor and those who are of a minority status including African American, Native American, Hispanic American, and other immigrant and non-English speaking population groups. Multiple studies of the impact of high heat index, extreme cold, hurricanes, flooding, and wildfires, find evidence that people of color, including Black, LatinX, Native American, Pacific Islander, and Asian communities are at a higher risk of climate-related health impacts and have more emergency department visits associated with extreme weather than those of the majority (17). These studies have also found evidence of racial disparities related to climatic changes with respect to mortality, respiratory and cardiovascular disease, mental health, and heat-related illness among adults (17).

Impact of Heat in Cities

Heat is among the deadliest weather-related phenomena in the US. The number of heat-related deaths is likely to increase with increasing ambient temperatures, particularly in urban areas (18). It is not surprising that poorer families tend to live in urban areas, especially in older and smaller houses or apartments and in more crowded areas with less green space. These urban neighborhoods tend to have a dense population of predominantly minority communities and have been referred to as ‘ghettos’. The historic roots of these neighborhoods lie in the 1930s, during the great depression, when the New Deal provided the suffering populace with jobs and loans to buy houses. The distribution of housing loans was managed by the Homeowners Loan Association, which divided city areas into four categories based on perceived likelihood to repay the loans. These designated areas were color-coded with the lowest risk being green and the highest risk being red. Loans were provided to people living in the three higher categories, but not to the lowest, which was red. This red area was inhabited predominantly by poor African American families, who, having been denied loans, continued to live in poverty without home equity and the opportunity to break out of a state of poverty. This phenomenon came to be known as 'Redlining' and is represented by the poor neighborhoods still occupied by poor people of color today (19). The built environment in these neighborhoods often consist of concrete and steel buildings and asphalt covered roads that reflect the sun's heat with no respite from the shade from trees, greenery, or bodies of water to absorb the heat and mitigate the temperature. As a result, these urban areas have a higher ambient temperature than surrounding suburban or peri-urban areas and are referred to as 'heat islands' (see Figure 3). These heat islands can have higher daytime temperatures up to 7°F higher than in outlying areas and less nighttime cooling with temperatures that can be up to 5°F higher (20,21). Therefore, people, and particularly children, who live in these urban 'heat islands' are more likely to experience increased ambient temperatures and consequently are more likely to suffer heat stress and heat related illnesses, with associated morbidity and mortality (21).

The extent to which the legacy of environmental injustice as a consequence of redlining with the confinement of poor African American and other minority populations in more urban settings with a greater heat index, can be seen in the juxtaposition of data on family income and ambient temperatures in various cities in the US (22). As an example of the relationship between income and heat exposure in a city, Figure 4 depicts the situation in Oakland California: the right sided image is of income, as depicted in green shading with the more intense green representing the higher income areas, while the image on the left is of the surface temperature with the higher temperatures being in the more intense red color. The higher the income, the lower the surface temperature, and the lower the income the higher the surface temperature (23). This juxtaposition dramatically demonstrates the relationship between low income and exposure to heat in the case of the poor community or relative coolness in the affluent neighborhoods.

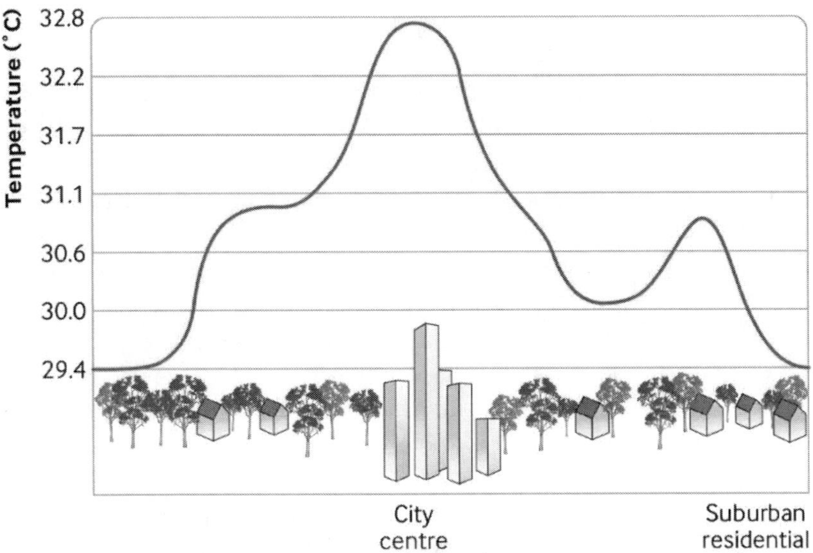

Figure 3. Urban heat island (21).

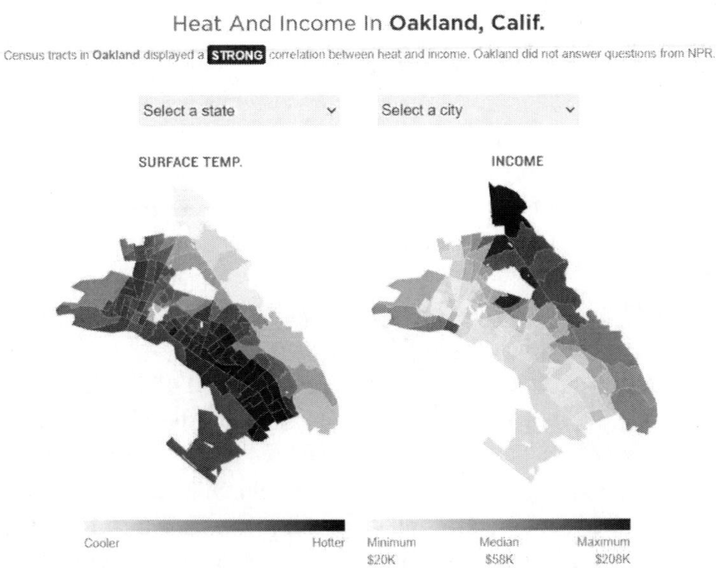

Figure 4. Comparison between income and heat exposure and experience within a city illustrating increased likelihood of heat exposure among people in poorer neighborhoods, compared to those in more affluent neighborhoods (23).

Heat and Drought in the Southwest United States

In precolonial times, the Indigenous peoples of the American continents, roamed or settled across the entire land mass with different tribes occupying specific areas of the country. Although it is difficult to precisely estimate the size of the precolonial population of Indigenous people in the continental US, estimates vary from an about 50 to over 100 million people (24). From 2017 to 2021, the US Census Bureau reports between 2.7 million U.S. residents identified as American Indian and Alaska Native alone, and 6.3 million as American Indian and Alaska Native alone or in combination with one or more other races (25).

With the advent of the European colonists from 1492 onwards, and the progressively increasing numbers of immigrants from Europe, battles for land between the European colonists and the Indigenous people were numerous and increasingly fierce (26). Initially the competition and conflicts were based in the eastern area of the continent where most of the colonists arrived by sea from European countries. These conflicts led progressively to decimation of local tribal communities and the forcible eviction of the remaining populations to the west of the Mississippi, effectively claiming exclusive rights to the entire land east of the Mississippi (26). One of these mass evictions became known as the Trail of Tears, when 17,000 Cherokee from the Southeast were forced to walk 1,200-miles to be resettled in Oklahoma. Approximately 6,000 men, women, and children are reported to have died on route (27). As a consequence of the resettlement process, most of the remaining Indigenous Tribal communities are located on small reservations, in relatively poor-quality land, in the western part of the US.

In early 2020, widespread drought developed in the US Southwest. Over the 20-month period from January 2020 through August 2021, the Southwest US received the lowest total precipitation and had the third-highest daily average temperatures since 1895 (28). The combined factors of record low precipitation and near-record high temperatures fueled the drought (29). This drought led to unprecedented water shortages in western reservoirs further threatening drinking, agricultural, and tribal water supplies, as well as fishing and recreational activities, and affecting electricity supply generated from hydroelectric plants. While all inhabitants in this region were impacted by the heat and drought, those with the least resources suffered the most, particularly those with the least access to reliable sources of water on a regular basis. The Navajo Nation is located on more than 27,000 square miles in the Southwest US within portions of Arizona, New Mexico, and Utah, with an on-reservation population is 173,667 out of a total Navajo population is 332,129 (30). Approximately thirty percent of the Navajo Nation population does not have access to clean reliable drinking water. In addition, they have limited access to water for their commercial businesses, health care needs, schools, and other facilities, and significantly, for irrigation of their crops and for their livestock, dramatically reducing and, in some cases, devastating their livelihood in agricultural productivity (31).

Furthermore, estimates suggest that approximately 15% of the Navajo population do not have access to piped water to their homes (31). Without piped water, residents are required to haul water either from regulated watering points or resort to water from unregulated sources, such as springs and windmill wells, designated for livestock. The number of unregulated water sources on the Navajo Nation is estimated to be in the thousands. The Navajo Nation Environmental Protection Agency (NNEPA) policy prohibits the use of unregulated water sources for human consumption because these sources are not routinely tested and regulated in accordance with the Safe Drinking Water Act (32). These unregulated water sources are susceptible to bacterial contamination, including fecal coliforms. Some of these unregulated water sources also exceed drinking water standards for contaminants such as uranium and other chemicals. Nevertheless, human consumption of unregulated water is reportedly widespread due to a lack of regulated water systems in the more remote and sparsely populated regions of the Navajo Nation. By some estimates, 48% of households on Indian reservations do not have clean water or adequate sanitation. Thus, the children are more likely to suffer from gastrointestinal disorders and the toxic effects from ingestion of water contaminated by chemicals (33). Indigenous communities in the US have suffered and continue to suffer inequities in access to clean water and other insecurities stemming from colonization and perpetuated by federal policy (34).

Little is understood about the long term, health consequences of drought, in particular, the implications for mental health as a consequence of loss of livelihood, diminished social support, and rupture of place bonds. We still have much to learn about the mental health implications of drought, and the inevitability of an increase in the risk of drought (35).

Another example of the lack of rainfall and excessive water usage has also led to the drying of the landlocked Salton Sea. The soil and dust that remained in what was the seabed contains toxins that had been dissolved in the sea water and now precipitated in the loose soil. Prevailing winds blow this toxic dust into surrounding residential areas causing significant respiratory disorders, especially among children and, significantly, children of the predominantly low-income, Mexican-American communities and families who live in close proximity to the source (36).

The drought has also contributed to the massive western wildfire seasons of 2020 and 2021, and the associated consequences on the environment and on human health. An estimated 7.4 million children in the US are affected by wildfire smoke each year causing significant respiratory pathology. In a study of 170,000 emergency and urgent care visits in San Diego from 2011-2017, emergency health care visits were 10x higher for air pollution from wildfire smoke than from other sources (37).

Climate Change and Global Health Disparities

Higher-than-average ambient temperatures in the Sahel of Africa have contributed to loss of vegetation and desertification significantly affecting agriculture and food supply contributing to food insecurity and malnutrition, particularly in children (38). Food insecurity, in turn, leads to migration to more promising areas for sustenance. These migrations result in destabilization of surrounding areas and may spark conflicts (39). These conflicts then exacerbate the degrading of the environment, with destruction and devastation of the environment and decimation of populations worsening the food insecurity and general sense of desperation further exacerbating migration (40). The data on migration shows that there are over 280 million migrants worldwide and many from Africa who mostly migrate to other countries in Africa, some of whom undertake desperate and grave risks in crossing the Mediterranean in unseaworthy vessels to refuge in European countries (41).

Recently, the civil war in South Sudan has compounded the impact of climate change and has aggravated food insecurity for more than seven million people with a projection by UNICEF that 1.4 million children will suffer malnutrition (42). The World Health Organization projects that there will be 77,000 to 131,000 additional deaths in children under five years from climate-related undernutrition in 2030, contributing to 45% of total deaths in children < 5 years old (43).

The World Bank explored the scenario of populations around the world who are forced to leave their homes because of climate related impacts on their lives and livelihoods, particularly in the agricultural areas (44). Many seek alternative living arrangements within their own countries, but many will seek refuge in more promising lands. The report states that most of the people affected by the impacts of climate change on their lives and livelihoods are likely to be from Africa and Asia, attested to by the boatloads of people leaving Africa to undertake often perilous journeys across the Mediterranean to find a more promising future in Europe. The World Bank also predicts that Latin America will be severely impacted by climate change, and project up to 3.9 million climate refugees by 2050. Although most will be internal refugees, many will seek asylum in the US, which will add stress to the politics and economics of the country and could result in major increases in political and economic disruption (44). Furthermore, the United Nations office on migration reports that 1 in every 8 migrants worldwide is a child representing the fastest growing segment of child and youth population in a number of high and middle-income countries across the world. These children are disproportionately vulnerable to violence, abuse, exploitation, trafficking, and detention, all of which are intensified for unaccompanied or separated migrant children (45).

According to the UN, nearly 10 million children were displaced as a result of weather-related events in 2020, and half the world's young people now live in regions at high risk from the impacts of climate change. In 2020 alone, nearly 10 million children were displaced in the aftermath of weather-related shocks. With around one billion children – nearly half of the world's 2.2 billion children – living in 33 countries at high risk of the impacts of climate change, suggesting that millions more children could be on the move in the coming years (46).

Challenges of Climate Change

Given the trend in global temperature rise, and the associated climate change with an increase in frequency and intensity of extreme weather events and disasters, we are witnessing increasing impacts on human habitation and human health, especially for the most vulnerable population groups which are children from poor and marginalized communities and countries. The accelerating pace of climate change and the continued vacillation and inaction of the world's leaders who currently hold the power to make the critical decisions, leaves the planet in a state of disruption that threatens the ecology of the planet to potentially massive devastation. Although there are organizations and agencies that are working overtime to make positive changes to halt, and possibly even reverse the trend, notably to reduce and potentially eliminate the use of fossil fuels that have contributed to this ongoing destructive process, we are all responsible for the state we are in and therefore, we are all responsible to make an effort to contribute to healing the planet and restoring a sense of balance and equity. As it is said in the Ethics of the Fathers: *although it is not your duty to complete the task, you are not at liberty to desist from undertaking the challenge* (47). In accepting this challenge, we have adapted our Break the Cycle of Children's Environmental Health Disparities program and paradigm to address Climate Change (48).

Breaking the Cycle of Climate Change

Conceptually, climate change can be viewed as a self-perpetuating cycle that needs to be broken to assure planetary health, ecological restoration, and human health (see Figure 5). In this publication we present a sample of the interdisciplinary array of creative work by students from universities in the US who participated in the 17th Annual *Break the Cycle* program in 2022 and focused on climate change and vulnerable populations. The three student projects in this section raise awareness of the historic impact of 'red-lining' on children's exposure to heat and pollution decades later, the phenomenon of eco-anxiety among youth and how to mitigate this by encouraging youth activism and the impact of climate change on healthcare among clinicians providing services to underserved populations. The other three papers represent perspectives on the impact of climate change on children's health disparities, the impact of climate change on migration, particularly from Latin America, and on climate change and climate equity from a colonial perspective.

Children and young people are least responsible for climate change, and yet are among the most vulnerable to its effects. When the generations that have created and perpetuated the extractive practices that caused global warming and climate change have disappeared, the planet will be left in a distressed state for the future generations. Today's children and young people are not only experiencing the impacts of climate change physically and physiologically, but they are also experiencing it psychologically and emotionally. Climate Change is a significant contributor to the increase in mental health disorders we are seeing in our society today (13, 49). In this volume, Nutkiewicz et al. explore the phenomenon of eco-anxiety and strategies to promote resilience through youth activism. It is our responsibility to listen to the children, to note their vulnerability and to heed the warnings and pleas for us to take action to arrest global warming.

As in so many situations, it is the children from poor communities and communities of color who suffer the most (48). Nenon et al. in this volume, explore the impact of climate change on communities, that were historically redlined, finding the disparities in exposure to higher ambient temperatures among other air pollutants that contribute to climate change. Berger at al also in this volume, note the increased vulnerability of children from poor communities and communities of color to a variety of climate change consequences, appealing for increased awareness and the need for climate action, focusing on youth to become informed and involved to help effect the necessary changes to mitigate climate change and promote climate equity and climate justice. Balaban et al. explore the provision of climate informed care among physicians providing health care to underserved communities and populations.

Figure 5. The cycle of climate change.

By the same token, populations and countries least responsible for the extractive practices that have resulted in overconsumption though fossil fuel use and over production of the greenhouse gas pollution that causes global warming and climate change, are suffering the most. This is clearly laid out by Faerron Guzmán et al. in their exploration and analysis also published in this volume. The impact of climate change on the more vulnerable countries which have not contributed significantly to climate change has resulted in disruption of their ecology and agricultural production affecting their lives and livelihood promoting migration. This migration not only puts migrants at risk, especially the children, but results in political disruptions in the destination countries which can have wide ranging political ramifications. In the paper by Tobarra et al., the migration from Latin America to North America is explored and some positive notes are sounded.

Historical Residential Redlining and Children's Vulnerability to Climate Change in the Southeast United States

Devon Nenon and Eric S Coker, BS, MS, PhD
Department of Environmental and Global Health, College of Public Health and Health Professions, University of Florida, Gainesville, Florida, United States of America

Abstract: Little is known about the impacts of historical redlining, a racially discriminatory housing practice, on future or current cumulative children's environmental health burdens. The Southeast US has one of the hottest climates in the United States (US) and the severity of extreme heat is expected to increase in the region due to climate change. The Southeast US also faces high levels of concentrated child poverty and air pollution-related inequalities. This study investigates relationships between redlining and combined stressors for multiple dimensions of the environment, including differences in climate change-related temperatures, in Alabama, Florida, Georgia and South Carolina. Methods: We combined data on historical ambient $PM_{2.5}$ and maximum temperature, climate change-related temperature projections, greenspace (normalized differential vegetative index [NDVI)) and child poverty. Multilevel linear regression and cluster analysis tested associations between Homeowners' Loan Corporation grades (A through D) and multiple environmental stressors, including differences in child poverty and climate change-related temperatures. Results: Relative to grade A neighborhoods

(Redlined neighborhoods), redlined neighborhoods are associated with higher historical maximum temperature (β = 0.07 [95% CI: 0.03, 0.11]), average temperatures in 2080 (β = 0.03 [0.00, 0.05]), summertime temperatures in 2080 (β = 0.06 [0.03, 0.09]), and changes in summertime temperatures (β = 0.006 [0.002, 0.01]). Average $PM_{2.5}$ (β = 0.24 [0.16, 0.32]) and NDVI (β = -0.06 [-0.08, -0.04]) are significantly worse in redlined neighborhoods. Cluster analysis identified redlined neighborhoods vulnerable to heat-related climate change impacts and combined high child poverty and environmental stressors. Conclusions: Historical redlining is associated with climate-related heat burdens in the Southeast US. This study can inform targeted climate adaptation efforts for communities impacted by structural racism.

Encouraging Youth Activism to Combat Climate Change and Reduce Eco-Anxiety

Lexi Nutkiewicz, MPH, Aminah Habib, BA, Miguel Duenas, MPH, Lorely Ruiz, MPH Gwendolyn Hoff Anderson, BA, and Kelly L'Engle, PhD, MPH
School of Nursing and Health Professions, University of San Francisco, San Francisco, California, United States of America

Abstract: Climate change is an ongoing public health crisis. Efforts to reverse climate change and protect future generations' physical and mental health are more urgent than ever. Youth are particularly vulnerable to a form of anxiety related to stress or distress caused by awareness of environmental changes, termed eco-anxiety. We conducted a scoping review to better understand the value of youth climate change activism as a strategy to combat climate change and mitigate eco-anxiety. Our review highlights risk factors for eco-anxiety among adolescents and gaps in the literature, considers the benefits and drawbacks of youth activism, and discusses models and strategies to promote youth engagement in climate change activism. To identify relevant literature, we searched a variety of databases, related research suggested by database algorithms, reference lists of resulting literature, and grey literature sources such as global reports and governmental resources. We summarize the findings on eco-anxiety and activism, and we provide recommendations for future research to inform effective public health practices and policy. As we shift our focus to the societal impacts of climate change, it is critical that we elevate the recognition of eco-anxiety as a reality and develop strategies to protect young people and their future.

A Survey to Inform Clinical Approaches to Address Climate Health and Equity

Eric Balaban, MD, Lesa Walker, MD, MPH
Department of Emergency Medicine, University of Colorado School of Medicine, Aurora, Colorado, United States of America and Environmental Defense Fund, United States Regional Office, Austin, Texas, United States of America

Abstract: Climate change is a present and growing threat to public health. Unfortunately, underrepresented populations experience a disproportionately increased risk to their health. This exploratory survey draws insights from United States-based healthcare providers who identify as part of and/or serve underrepresented communities. The survey assesses their clinical perspectives and practices in addressing the health impacts of climate change (climate-health) and associated health inequities. The survey was designed using input from healthcare leaders with expertise and experience serving underrepresented communities. It was disseminated via partnering clinical provider networks. Results were collated within a SurveyMonkey platform and analyzed using descriptive statistics. The surveyed medical providers recognized climate change and its potential impacts on patient health but were least familiar with its impacts on health inequities. Only 26% routinely screened for climate change-related health risks, and only 20% discussed these risks in clinical encounters. The most common barriers to screening were time constraints (72%) and competing priorities (53%). Participants suggested that a screening tool could be optimized by prioritizing expediency of use (70%) and incorporation into the electronic medical record system (67%). Conversing with patients about climate change and their health was most hindered by time constraints (68%) and the lack of familiarity with solutions to offer (53%). This study identifies opportunities to optimize educational, material, and systems-level interventions to expand climate-conscious clinical practice. Importantly, these results focus on the lived experiences of frontline providers to inform solutions.

Implications of Climate Change on Children's Health – An Equity Perspective

Martha Berger, MPA, Nse Obot Witherspoon, MPH, Justin Harris, and I Leslie Rubin, MD
Climate Equity Collaborative, Children's Environmental Health Network, Break the Cycle of Health Disparities, Inc.

Abstract: Climate change is a significant threat to human health as greenhouse gases accumulate, temperatures rise, sea levels rise, and extreme weather events become more intense and frequent. In communities throughout the world, those most often and most severely affected by climate change are those with the least means to protect themselves and to recover when disaster strikes, most notably children. The physical and mental health implications of climate change for children, from before birth, through infancy, during early childhood, and as teenagers, are distinct in the short- and long-term health effects of environmental exposures. For children in under-resourced communities, where adverse environmental exposures are more likely, and the health effects are disproportionately more harmful. Furthermore, the current trajectory of climate change is creating both anxiety and positive energy in the youth of today to become part of advocating for and making positive changes to address climate change and improve their physical and mental health. In addition, concerned citizens are forging new alliances and partnerships to address the impacts of climate change on children's health. There is an urgent need to increase awareness and knowledge among health care professionals and the general public about climate change and its disproportionate impact on children, particularly on children in poverty and marginalized communities, and to develop collaborative strategies to combat climate change, protect the children and their future.

Understanding Climate Migration with a Focus on Latin American Children and Families

Esther Tobarra, MD, Fernando Wang, MSc and Leslie Rubin, MD
Cwm Taf Morgannwg University Health Board, Royal Glamorgan Hospital, Rhondda Cynon Taf, Wales, United Kingdom, Oncology Institute of Marseille, Marseille, France, Interuniversity Master in Planetary Health,

Universitat Oberta de Catalunya, ISGlobal, Pompeu Fabra, Barcelona, Spain

Abstract: Consequences of climate change are forcing an increasing number of people around the world to migrate from their climate-stressed countries of origin to countries that offer better access to vital resources and greater security. This trend can significantly impact migrant's health and socioeconomic outlooks, as well as having unanticipated implications for the countries of origin and destination. Many migrants are families with children, who are most vulnerable to the impacts of environmental hazards. Migration can disturb a child's formative years, increasing the risks of physical harm, mental illness, and social vulnerabilities, and, when traveling alone, becoming a target for kidnapping and human trafficking. We explore the example of an increasing number of migrants leaving countries in Latin America for neighbouring countries including the United States (US). Around half of those arriving to the US originate from the Central Dry Corridor of the Northern Triangle, with food insecurity being one of the primary reasons. There is an urgent need to protect the rights of this group of vulnerable children, ensuring their appropriate settlement, with access to healthcare and education. Health professionals can play an essential role in identifying risks to migrant children's physical and psychological health and advocating on their behalf. Social Prescribing is an example of a valuable strategy that offers a community-centred framework, favouring integration and personalised care, and taking a holistic approach to health and wellbeing. We also recognize the need for more research and information on optimal strategies to promote migrants' family wellbeing in their new country.

Acting on Power and Health Inequities: An Essential Approach to Global Environmental Challenges

Carlos A Faerron Guzmán, MD, MSc and Olena Zotova
University of Maryland, Baltimore, MD, United States of America, Planetary Health Alliance, Harvard TH Chan School of Public Health, Boston, Massachusetts, United States of America, ISGlobal, Barcelona, Spain and Pompeu Fabra University, Spain

Abstract: Human civilization has less than ten years remaining to limit environmental changes that risk undoing a century of hard-earned gains in global health and human wellbeing. Current actions are insufficient to address the health emergency caused by global environmental change. We provide a diagrammatic intersectional analysis of the interaction between power, health, and environmental change that explains the prevailing status quo in environmental action, despite the health losses experienced by "lower power" groups. Using real-world examples, we demonstrate how the power-health-environment nexus underlies current delays in redressing health and environmental inequities and how we must act on this nexus to protect global and planetary health from anthropogenic disruptions in Earth's natural systems.

Conclusion

The consequences of poverty and race on the health and well-being of children are compounded and amplified by the impact of climate change in the form of heat, drought, flooding, and extreme weather events. Poor children and children of color currently suffer disproportionately from the social, economic, environmental, and political realities in which they live, and are at greater risk for morbidity and mortality. The challenges are complex and multifaceted, requiring creative and coordinated responses at multiple levels. There are two fundamental challenges of undeniably gargantuan proportion, these are (a) the ever-increasing rise in global temperature without a sign of respite that requires a concerted effort to dramatically reduce the burning of fossil fuels, and (b) the intractable historic social, economic, and racial inequities, locally and globally that requires a concerted effort to confront historic injustices and develop strategies to promote justice and equity for all. There is no one solution to the challenges, but a global, concerted effort to tackle the different aspects in as many ways as possible by as many different large governmental entities, intermediate size public and private agencies, and organizations, as well as individuals and small groups.

We offer the concept of breaking the *Cycle of Health Disparities* and the *Cycle of Climate Change* as a framework for inspiring and encouraging the creative ideas and energy of our youth, to put their thoughts and ideas to work, and mobilize critical resources to support them in taking on these mammoth challenges. The future of the planet and life on it depends on the action we take today. The health and well-being of our children and their children hang in the balance, not in isolation, but in concert with our fellow human beings from different cultures and different parts of the world to work together and to preserve the fragile ecology of our planet which supports all life and nurtures resilience.

To this end, our *Break the Cycle Program* promotes awareness of the social, economic, environmental, and political determinants of health among our youth of today and encourages them to develop creative and innovative strategies to reduce and eliminate health disparities in our society, for the good of all. Furthermore, *Break the Cycle of Climate Change* raises awareness and stimulates the generation of projects that will translate into action in reducing the impact of climate change, and promoting resilience among young people of today who will be the leaders of tomorrow (48).

Participating students gain valuable knowledge and skills towards becoming the next generation of leaders, and, building on their knowledge and experience, to further the goals of reducing children's environmental health disparities, promoting health equity for all children, developing effective strategies to confront climate change, and championing climate resilience and social justice. In the process, we all gain knowledge, understanding and perspective to become more inclusive and open to innovations and progress.

Acknowledgments

Thanks to the additional editors of some of the papers who have given of their time and expertise, much appreciated especially our trusted friend and star guest editor, editor David Ervin. This publication was supported by the cooperative agreement award number 5 NU61TS000237-05 from the Agency for Toxic Substances and Disease Registry (ATSDR). Its contents are the responsibility of the authors and do not necessarily represent the official views of the Agency for Toxic Substances and Disease Registry (ATSDR). The US Environmental Protection Agency (EPA) supports the Pediatric Environmental Health Specialty Units (PEHSU) by providing partial funding to ATSDR under Inter-Agency Agreement number DW-75-95877701. Neither EPA nor ATSDR endorse the purchase of any commercial products or services mentioned in PEHSU publications.

References

[1] National Geographic. Industrialization, Labor, and Life. URL: https://education.nationalgeographic.org/resource/industrialization-labor-and-life/.

[2] Rubin IL. Break the cycle of environmental degradation: A commentary. Int Public Health J 2018;10(4):379-88.

[3] NASA. Global climate change. URL: https://climate.nasa.gov/global-warming-vs-climate-change/.

[4] Crimmins, A. Balbus J, Gamble JL, Beard CB, Bell JE, Dodgen D, et al. The impacts of climate change on human health in the United States: A scientific assessment. Washington, DC: US. Global Change Research Program, 2016. URL: http://dx.doi.org/10.7930/J0R49NQX.

[5] Lindsey R, Dahlan L. Climate change: Global temperature. URL: https://www.climate.gov/news-features/understanding-climate/climate-change-global-temperature#:~:text=Earth's%20temperature%20has%20risen%20 by,0.18%C2%B0%20C)%20per%20decade.

[6] NOAA. How is sea level rise related to climate change? URL: https://oceanservice.noaa.gov/facts/sealevelclimate.html.

[7] UNICEF. Devastating floods in Pakistan. URL: https://www.unicef.org/emergencies/devastating-floods-pakistan-2022.

[8] NOAA. Billion-dollar weather and climate disasters. URL: https://www.ncei.noaa.gov/monitoring-content/billions/images/2022-billion-dollar-disaster-map.png.

[9] Dennis B. The toll extreme weather took in the US during 2022, by the numbers. Washington Post. URL: https://www.washingtonpost.com/climate-environment/2022/12/30/blizzard-hurricane-drought-flood-tornado-2022.

[10] Xu Z, Etzel RA, Su H, Huang C, Guo Y, Tong S. Impact of ambient temperature on children's health: A systematic review. Environ Res 2012;117:120-31. doi: 10.1016/j.envres.2012.07.002.

[11] Sheffield PE, Herrera MT, Kinnee EJ, Clougherty JE. Not so little differences: Variation in hot weather risk to young children in New York City. Public Health 2018;161:119-26.

[12] Bernstein AS, Sun S, Weinberger KR, Spangler KR, Sheffield PE, Wellenius GA. Warm season and emergency department visits to US Children's Hospitals. Environ Health Perspect 2022;130(1):17001. doi: 10.1289/EHP8083.

[13] EPA. Climate change and children's health and well-being in the United States. Washington, DC: US Environmental Protection Agency, EPA 430-R-23-001, 2023.

[14] Vergunst F, Berry HL. Climate change and children's mental health: A developmental perspective. Clin Psychol Sci 2022;10(4):767-85. doi: 10.1177/21677026211040787.

[15] CDC. Caring for children in a disaster. URL: https://www.cdc.gov/childrenindisasters/features/disasters-mental-health.html#:~:text=After%20a%20 disaster%2C%20children%20may,can%20cause%20stress%20for%20families.

[16] Le Brocque R, De Young A, Montague G, Pocock S, March S, Triggell N, et al. Schools and natural disaster recovery: The unique and vital role that teachers and education professionals play in ensuring the mental health of students following natural disasters. J Psychol Couns Schools 2017;27(1):1-23. doi:10.1017/jgc.2016.17.

[17] Berberian AG, Gonzalez DJX, Cushing LJ. Racial disparities in climate change-related health effects in the United States. Curr Environ Health Rep 2022;9(3):451-64. doi: 10.1007/s40572-022-00360-w.

[18] Petkova EP, Bader DA, Anderson GB, Horton RM, Knowlton K, Kinney PL. Heat-related mortality in a warming climate: Projections for 12 US cities Int J Environ Res Public Health 2014;11:11371-83. doi:10.3390/ijerph111111371.

[19] Swope CB, Hernández D, Cushing LJ. The relationship of historical redlining with present-day neighborhood environmental and health outcomes: A scoping review and conceptual model. J Urban Health 2022;99(6):959-83. doi: 10.1007/s11524-022-00665-z.

[20] EPA. Learn about Heat Islands. URL: https://www.epa.gov/heatislands/learn-about-heat-islands#characteristics.

[21] Tong S, Prior J, McGregor G, Shi X, Kinney P. Urban heat: An increasing threat to global health. BMJ 2021;375:n2467. doi:10.1136/bmj.n2467.

[22] Mitchell BC, Chakraborty J. Landscapes of thermal inequity: Disproportionate exposure to urban heat in the three largest US cities. Environ Res Lett 2015;10(11):115005.

[23] Anderson M, McMinn S. As rising heat bakes US cities, the poor often feel it most. URL: https://www.npr.org/2019/09/03/754044732/as-rising-heat-bakes-u-s-cities-the-poor-often-feel-it-most.

[24] Denevan WM, ed. The native population of the Americas in 1492, Second revised edition. Madison, WI: University of Wisconsin Press, 1992.

[25] US Census Bureau. Census Bureau releases new American community survey selected population tables and American Indian and Alaska native tables, 2023 Jun 15. URL: https://www.census.gov/newsroom/press-releases/2023/acs-selected-population-aian-tables.html.

[26] Dunbar-Ortiz R. An indigenous peoples' history of the United States. Boston, MA: Beacon Press, 2014.

[27] NIH NLM Indian voices 1838: Cherokee die on Trail of Tears. URL: https://www.nlm.nih.gov/nativevoices/timeline/296.html#:~:text=The%20U.S.%20Department%20of%20War,called%20the%20Trail%20of%20Tears.

[28] NOAA/NIDIS. New NOAA Report: Exceptional southwest drought exacerbated by human-caused warming. URL: https://www.drought.gov/news/new-noaa-report-exceptional-southwest-drought-exacerbated-human-caused-warming.

[29] Mankin JS, Simpson I, Hoell A, Fu R, Lisonbee J, Sheffield A, Barrie D. NOAA drought task force report on the 2020–2021 Southwestern US drought. Washington, DC: NOAA Drought Task Force, MAPP, and NIDIS, 2021.

[30] Navajo Division of Health Navajo Epidemiology Center. Navajo population profile 2010. US Census. 2013. URL: https://nec.navajo-nsn.gov/Portals/0/Reports/NN2010PopulationProfile.pdf.

[31] Navajo Nation Department of Water Resources. URL: https://nndwr.navajo-nsn.gov/.

[32] EPA. Providing safe drinking water. URL: https://www.epa.gov/navajo-nation-uranium-cleanup/providing-safe-drinking-water-areas-abandoned-uranium-minesn.

[33] Hoover JH, Erdei E, Begay D, Gonzales M, Jarrett JM, Cheng PY, et al. Exposure to uranium and co-occurring metals among pregnant Navajo women. Environ Res 2020;190:109943. doi: 10.1016/j.envres.2020.109943.

[34] Tanana H, Combs J, Hoss A. Water is life: Law, systemic racism, and water security in Indian Country. Health Secur 2021;19(S1):S78-82. http://doi.org/10.1089/hs.2021.003.

[35] Vins H, Bell J, Saha S, Hess JJ. The mental health outcomes of drought: A systematic review and causal process diagram. Int J Environ Res Public Health 2015;12(10):13251-75. doi: 10.3390/ijerph121013251.

[36] Johnston JE, Razafy M, Lugo H, Olmedo L, Farzan SF. The disappearing Salton Sea: A critical reflection on the emerging environmental threat of disappearing saline lakes and potential impacts on children's health. Sci Total Environ 2019;663:804-17. doi: 10.1016/j.scitotenv.2019.01.365.

[37] Aguilera R, Corringham T, Gershunov A, Leibel S, Benmarhnia T. Fine particles in wildfire smoke and pediatric respiratory health in California. Pediatrics 2021;147(4): e2020027128.

[38] Baptista D, Farid M, Fayad D, Kemoe L, Lanci L, Mitra P, et al. Climate change and chronic food insecurity in Sub-Saharan Africa. Washington, DC: International Monetary Fund, 2022.

[39] UN. Conflict and climate. URL: https://unfccc.int/blog/conflict-and-climate.

[40] Sadiddin A, Cattaneo A, Cirillo M, Miller M. Food insecurity as a determinant of international migration: Evidence from Sub-Saharan Africa. Food Sec 2019;11: 515–30.

[41] IOM UN. Migration world migration report 2022. URL: https://worldmigrationreport.iom.int/wmr-2022-interactive/.

[42] UNICEF. Hunger and malnutrition being driven by climate crisis and conflict in South Sudan. URL: https://www.unicef.org/press-releases/hunger-and-malnutrition-being-driven-climate-crisis-and-conflict-south-sudan.

[43] Lloyd S, Kovats S, Chalabi Z. Undernutrition. In: Hales S, Kovats S, Lloyd S, Cambpell-Lendrum D, eds. Quantitative risk assessment of the effects of climate change on select causes of death, 2030s and 2050s. Geneva: World Health Organization, 2014:69–94.

[44] World Bank. Preparing for internal climate migration. Washington, DC: World Bank, Groundswell Policy Note 3, 2018. URL: https://documents1.worldbank.org/curated/en/983921522304806221/pdf/124724-BRI-PUBLIC-NEWSERIES-Groundswell-note-PN3.pdf.

[45] IOM UN Migration. Protection of children in migration. Brussels: IOM, 2019. URL: https://eea.iom.int/protection-children-migration.

[46] UNICEF. New guidelines provide first global policy framework on protecting children on the move in face of climate change. URL: https://www.unicef.org/press-releases/new-guidelines-provide-first-global-policy-framework-protect-children-move-displaced-face.

[47] Pirkei Avot. Ethics of the fathers. 2:16.

[48] Rubin IL, Witherspoon NO. Climate change, environmental justice and children's health: Break the cycle of climate change by cultivating future leaders. J Appl Res Child 2021;12(1):7.

[49] Vergunst F, Berry HL. Climate change and children's mental health: A developmental perspective. Clin Psychol Sci 2022;10(4):767-85. doi: 10.1177/21677026211040787. Epub 2021 Sep 14. PMID: 35846172; PMCID: PMC9280699.

Section One: Break the Cycle

Chapter 2

Encouraging Youth Activism to Combat Climate Change and Reduce Eco-Anxiety

Lexi Nutkiewicz, MPH
Aminah Habib, MPH
Miguel Duenas, MPH
Lorely Ruiz, MPH
Gwendolyn Hoff Anderson, MPH
and Kelly L'Engle*, PhD, MPH
School of Nursing and Health Professions, University of San Francisco, San Francisco, California, United States of America

Abstract

Climate change is an ongoing public health crisis. Efforts to reverse climate change and protect future generations' physical and mental health are more urgent than ever. Youth are particularly vulnerable to a form of anxiety related to stress or distress caused by awareness of environmental changes, termed eco-anxiety. We conducted a scoping review to better understand the value of youth climate change activism as a strategy to combat climate change and mitigate eco-anxiety. Our review highlights risk factors for eco-anxiety among adolescents and gaps in the literature, considers the benefits and drawbacks of youth activism, and discusses models and strategies to promote youth engagement in climate change activism. To identify relevant literature,

* ***Correspondence:*** Kelly L'Engle, School of Nursing and Health Professions, University of San Francisco, 2130 Fulton Street, San Francisco, 94117, United States.
Email: klengle@usfca.edu.

In: Environmental Health: Leaders for Climate Change
Editors: I Leslie Rubin and Joav Merrick
ISBN: 979-8-89530-347-4

we searched a variety of databases, related research suggested by database algorithms, reference lists of resulting literature, and grey literature sources such as global reports and governmental resources. We summarize the findings on eco-anxiety and activism, and we provide recommendations for future research to inform effective public health practices and policy. As we shift our focus to the societal impacts of climate change, it is critical that we elevate the recognition of eco-anxiety as a reality and develop strategies to protect young people and their future.

Introduction

Climate change is an ongoing public health crisis. With a growing global population, increased industrialization, and heightened consumerism, anthropogenic climate change is dramatically impacting human health and well-being. The Intergovernmental Panel on Climate Change (IPCC) report cautions that climate change is expected to have adverse and often severe impacts on the well-being of children and adolescents (1). Efforts to protect the physical and mental health of future generations are more urgent than ever.

Climate change affects the social and environmental determinants of health by disrupting the basic requirements needed to live a healthy life such as clean water, air, and access to food. According to the World Health Organization, youth are particularly vulnerable to the impacts of climate change, and it is estimated that they will suffer more than 80% of illnesses, injuries, and deaths due to climate change (2). Beyond the biological and psychological impacts of climate change, social, economic, and ecological factors that affect health will also be disrupted since many of these factors intersect and interact. For example, rising global temperatures affect the water cycle and influence rainfall patterns, which impacts crop production, water contamination, and sea level rise. This is just one illustration of how climate change will affect the global economy and other social, economic, and environmental determinants of health. Even countries that have not experienced drastic direct effects of climate change will face scarcity as critical natural resources become strained to meet global demand.

The economic burdens of climate change differ based on country and geographic location, and poorer countries in the tropics and Southern Hemisphere subtropics may experience the most significant detrimental economic impacts from the climate crisis (3). About 85% of children in the world live in low to middle-income countries, which are generally located in geographic areas with weak infrastructure and sparse resources to combat climate change impacts (2). This puts the majority of youth at a further economic disadvantage to fight the present and future effects of climate change, which also have a significant negative impact on their well-being.

Researchers have consistently found that climate change impacts adolescents' mental health in unique and complex ways (4-6). An example of climate-related effects that have been well-researched for their impact on adolescents is the damaging mental health effects of severe weather events. The aftermath of severe weather events can adversely alter the social determinants of health, resulting in increased poverty, less stable housing, reduced caregiver presence, and interrupted access to education. Children and adolescents are especially vulnerable to post-traumatic stress after an extreme weather event. Research has found that the number of children with significant post-traumatic stress disorder (PTSD) symptoms can be as high as 92% during the subacute phase post-wildfire (7). Extreme weather events are associated with psychological distress, poor mental health (especially in people with pre-existing mental health conditions), increased psychiatric hospitalizations, and heightened suicide rates (8). In recent years, the term eco-anxiety has gained popularity among media and researchers to describe the distress and poor mental health outcomes people experience that are related to climate change and the environment (9). However, eco-anxiety among youth and treatment options to alleviate the symptoms remains a nascent field of research, despite the increased vulnerability of youth to the mental health effects of climate change.

As the climate crisis persists, youth are becoming more uncertain about their future and are expressing their fear through global youth climate strikes. Greta Thunberg's Fridays for Future ("Skolstrejk för klimatet") movement mobilized millions of youth climate activists to engage in collective climate change action, with over 10 million participants, according to best estimates (10). The youth uprising against climate change is no surprise, as researchers have found that worrying about climate change can lead to powerful action among young people (11). To better understand the relationship between eco-anxiety and action, and to explore the value of climate change activism as a potential mitigation strategy for symptoms of eco-anxiety, we conducted a scoping review to explore the benefits and drawbacks of youth activism, youth motivations for activism, and models and strategies to promote engagement in climate activism. The findings will inform suggestions for improving youth well-being through research, practice, and policy recommendations.

Review

A scoping review synthesizes available evidence and concepts related to a defined area to answer exploratory research questions, rather than responding to a specific research question like a systematic review (12). A scoping review can also include diverse literature sources, including research from other disciplines and grey literature (12). Because the mental health impacts of climate change and environmental activism among youth are under-researched areas, we chose to conduct a scoping review to explore current literature and emerging programs. To conduct our scoping review, we followed the structure compiled by Gutirrez-Bucheli et al. (12) and started by forming research questions aligned with our purpose. The following research questions drove our review:

1. How does climate change affect the mental health of youth?
2. What motivates youth to engage with activism and how does it affect them?
3. How can climate change activism be applied to address eco-anxiety among youth?
4. What actionable steps can be taken to address eco-anxiety through climate change activism?

To identify relevant literature, we searched a variety of databases, related research suggested by database algorithms, reference lists of resulting literature, and grey literature sources such as global reports and governmental resources. Search engines, search terms, and eligibility criteria are provided in the appendix. Initial search terms were chosen based on our research questions and were expanded to include keywords from the initial literature searches. To better understand the scope of the term "eco-anxiety," we focused specifically on this term and omitted similar terms such as "solastalgia" and "eco-angst." Articles with both qualitative and quantitative data were included. Although the focus of the review was on North American youth, literature that discussed youth of other nationalities and adults who either were affected by climate change or participated in activism were included if literature specific to our chosen population was limited or absent. We focused on literature published after 2010 to account for the growing impact of climate change, new research on adolescent development, and the use of social media as a form of activism and information sharing in the past decade.

Titles and abstracts of identified literature were reviewed to determine if they contributed to our driving research questions. The selected literature was then reviewed and charted onto spreadsheets shared amongst the research team on Google Sheets. Literature was sorted into the following categories: Climate Change and Well-Being, Eco-Anxiety and Activism, Activism among Youth, Climate Change Activism and Encouragement, Communication, Reviews, and News Articles. Within each sheet, the following information was extracted from each piece of literature: Year and Journal of Publication, Study Focus, Study Design, Population, Setting, Results, Discussion, and additional commentary. Additional critical review of the resulting data determined inclusion in this scoping review.

Findings

With growing awareness of the impacts of climate change on mental health, researchers have coined the term "eco-anxiety" to help understand the connection between climate change and anxiety about the climate crisis. Eco-anxiety has been defined in multiple ways; Table 1 presents the varied definitions identified after a thorough literature search of peer-reviewed articles focused specifically on eco-anxiety. Most scholars agree that eco-anxiety is an emotional response to the current state of the environment. While the emotional response to climate change is broadly defined, one of the main differences between definitions is the *severity* of the emotional response. Descriptions of severity range from less severe phrases such as "non-specific worry" and "generalized sense" to more severe phrases such as "chronic fear," "severe and debilitating worry," and "paralyzing powerlessness." Additionally, most scholars acknowledge that eco-anxiety can arise from direct experiences of climate change but also from indirect knowledge or anticipation of climate change and the climate crisis.

The adverse emotional and behavioral symptoms of eco-anxiety include sadness, depression, irritability, feelings of uncertainty, frustration, panic attacks, loss of appetite, and hopelessness (9, 24). However, there is debate about whether eco-anxiety is truly an anxiety disorder or simply a justifiable reaction to the climate crisis. Eco-anxiety shares similar characteristics with the Diagnostic and statistical manual of mental disorders - Fifth edition (DSM-5) definition of generalized anxiety disorder (GAD).

Table 1. Operationalization of "eco-anxiety" in current literature

Author (Year)	Operationalization of eco-anxiety
Albrecht (13)	Generalized sense that the ecological foundations of existence are in the process of collapse
Albrect (13), as cited by Pihkala (14)	Non-specific worry about our relationship to support environments
American Psychological Association and ecoAmerica, as cited by Schreiber (15)	A chronic fear of environmental doom
Clayton (16)	Anxiety associated with perceptions about climate change, even among people who have not experienced the direct impacts.
Clayton and Karazsia (17)	Form of negative emotional response to climate change
Comtesse et al. (18)	Response to impending threats of climate change
Doherty and Clayton (19)	Indirect, vicarious impacts of climate change that include intense emotions associated with observation of climate change effects worldwide and anxiety and uncertainty about the unprecedented scale of current and future risks to humans and other species
Gifford and Gifford (20)	Severe and debilitating worry about the risk that may be insignificant and is not associated with the more proactive behavior associated with habitual ecological worrying
Hickman (21)	Distress relating to the climate and ecological crises
Lertzman, as cited by Dockett (22)	Form of pre-traumatic stress disorder in which traumatic consequences are anticipated and felt before the event takes place
Pihkala (14,23)	A contemporary form of the phenomenon where the state of the world affected a person's emotional well-being
Schwartz et al. (24)	Negative cognitive, emotional, and behavioral responses associated with concerns about climate change
Stanley et al. (25)	Anxiety experienced in response to the ecological crisis
Usher et al. (26)	Specific form of anxiety relating to stress or distress caused by environmental changes and our knowledge of them
Van Susteren, as cited by Baudon and Jachens (27)	Environmental melancholia - a combined sense of primal loss and paralyzing powerlessness

For example, Schwartz and colleagues (24) found that functional impairment and cognitive-emotional impairment associated with climate change anxiety were significantly associated with GAD symptoms. Even though these data were collected from adults, they indicate that GAD symptoms can appear in climate change-induced anxiety. From the various definitions listed in Table 1, we believe Usher best encapsulates the symptoms and experiences of eco-anxiety which is specific to the climate crisis and describes how climate change manifests itself in our physical environment and everyday lives (26). From a behavioral health perspective, recognizing a person's behaviors and attitudes towards an idea or issue is how we can better understand their concerns and work towards appropriate treatment options.

In studying eco-anxiety, we must examine the various intersectional characteristics that can make someone more susceptible to eco-anxiety. Most eco-anxiety research has focused on adults, but given the multiple and potentially severe impacts of the climate crisis on young people, eco-anxiety should not be overlooked as a mental health concern for adolescents.

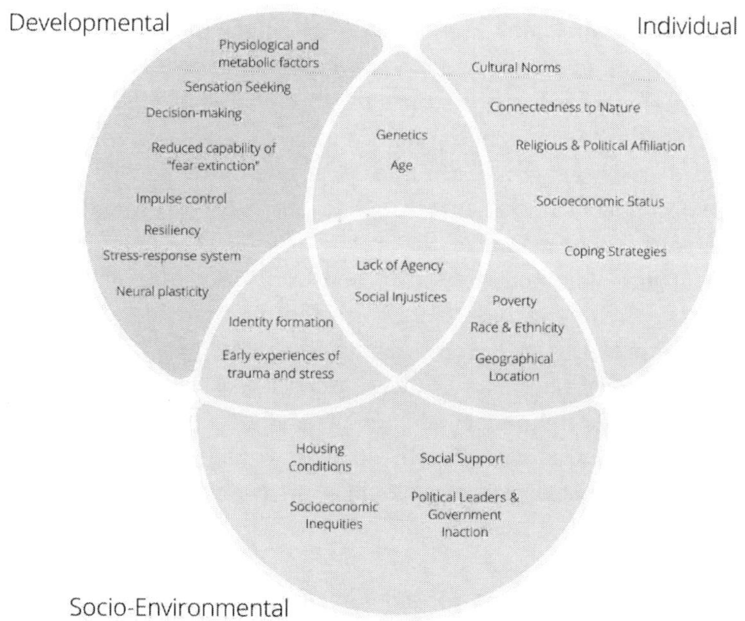

Figure 1. Developmental, individual, and socio-environmental factors that influence the health impacts of climate change on adolescents.

The Disproportionate Effects of Climate Change on Adolescents' Mental Health

Adolescents are disproportionately impacted by climate change due to their critical developmental stage and lack of power and control over their environment (6, 28). These factors intersect with other developmental considerations, individual factors, and socio-environmental influences that place adolescents at a greater risk of mental health concerns (see Figure 1).

Developmental Factors

Adolescence is a unique period of development where youth experience the greatest physical, mental, and social growth and change after infancy (29, 30). Physiological and metabolic factors can make young people more vulnerable to the direct effects of climate change than adults. Children and teenagers, especially infants and small children, are less able to thermoregulate, and therefore less adaptable to climate-related exposures and climate variability compared to adults (31). Neurologists have noted that adolescent brains are more powerful than adults and provide an incredible capacity to learn. However, they are less protected against stress compared to adults, because their different levels and responses to stress hormones make them less able to manage emotions and more prone to respond with extreme emotions, especially in crisis situations (30). Prior exposure to trauma and stress, such as adverse childhood experiences such as physical or emotional abuse, systemic racism, or climate disaster events, may further impair adolescents' ability to manage fear and stress (30) and lead to learning and behavioral problems (4). Anxiety disorders often emerge during adolescence, with an estimated 2-9% of US adolescents having some type of anxiety order, with girls having higher rates and earlier onset (30). Thus, there is a strong connection between environmental stressors and anxiety, and developmental characteristics during adolescence that increase young people's susceptibility to both anxiety and the effects of the climate crisis (32).

Individual Factors

Individual-level risk factors for adolescent eco-anxiety include genetic predisposition, pre-existing conditions, various demographic factors, and connection to the natural world (see Figure 1). Genetic predisposition to mental health disorders such as PTSD, depression, and anxiety has been linked to children's susceptibility to developing these conditions following a natural disaster (33). La Greca and colleagues (33) found that children displaying higher genetic markers of brain-derived neurotrophic factor (BDNF) reported greater symptoms of PTSD after hurricane exposure or hurricane-related stressors. For adolescents with pre-existing conditions, climate change and extreme weather changes can impact the effectiveness of psychiatric medications, as well as make many chronic conditions more difficult to manage (5). Many psychiatric medications affect the body's ability to thermoregulate and place adolescents at an increased risk of hyperthermia (5). While little is known about the trends and patterns of antipsychotic medication use among young people in the United States, data from 2012 revealed that 0.14% of UK (United Kingdom) adolescents and 1.03% of adolescents in the Netherlands were prescribed antipsychotics (5). Demographic factors including but not limited to socioeconomic status (SES), age, race and ethnicity, religious and political ideologies, cultural norms, connectedness to nature, and education can affect the onset and severity of eco-anxiety (6, 34-37). Climate-related disasters and severe weather patterns can cause disproportionate economic stress on low-income individuals because of inadequate infrastructure, lack of proper insurance, and limited access to resources (8, 35). More specifically, the climate crisis disproportionately impacts low-income and marginalized youth due to societal disadvantages including but not limited to access to healthcare, proximity to environmentally dangerous structures, such as landfills, industrial sites, and freeways, with fewer green spaces. Sachdeva (36) notes that religion can shape individual environmental concerns and actions. Lastly, people who care more about environmental issues and/or feel a deeper connection to the natural world, notably young people, and Indigenous groups, tend to be more vulnerable to developing eco-anxiety (9, 17).

Socio-Environmental Factors

Socio-environmental factors also have an important role in determining the prevalence of eco-anxiety in adolescents. The geographical location of their home can greatly increase stress and anxiety from direct exposure to climate change-related disasters, pollution, and extreme temperatures. Location of residence intersects with an adolescent's socioeconomic status and race/ethnicity because studies reveal that communities of color and low-income populations are more likely to live in areas with higher pollution and warmer temperatures (38). Homes in low-income neighborhoods may also lack proper infrastructure, which is more prone to destruction during climate-related events. Furthermore, low-income communities are more likely to live closer to industrial facilities, highways, and electric power grids, placing them at higher risk of exposure to exhaust and toxic chemical spills and leaks (35).

Another key component of an adolescent's environment is their social support network. Social support contributes to an adolescent's ability to build resilience and cope with climate change (6). Adolescents rely heavily on their social environment for emotional stability. When a climate-related event occurs, young people can be exposed to family stress and displacement, putting them at risk for developing mental health concerns such as PTSD, anxiety, sleep disorders, and substance abuse (4).

Societal injustices aggravate and perpetuate exposure to climate change-related events and mental health conditions. Systemic inequities have disproportionately placed communities of color in undesirable locations with limited access to viable resources. The cumulative impact of individual social determinants of health such as poverty and racism can amplify the negative effects of climate change. According to Ragavan, Marcil and Garg (39), poverty disproportionately affects children, communities of color, and communities of lower socioeconomic status. Climate change is more likely to impact people experiencing poverty and can also further exacerbate poverty. It is, therefore, critical that we urgently address climate change to help break the cycle of environmental health disparities in marginalized communities.

Young people are notably disadvantaged because of their age and the powerlessness associated with their minor status. Adolescents have to rely on adults for decisions about climate change and resources to protect them against its effects (28). Hickman and colleagues (40) conducted a global survey of 10,000 young people (ages 16 to 25 years) and found that young people were most distressed when they believed the government's response to climate change was inadequate. Because youth are at grater risk for eco-anxiety and other negative health effects from the climate crisis, as well as live with the effects of climate change for longer, there is a pressing need for youth-focused mental health resources, especially resources that focus on climate-related mental health concerns. However, eco-anxiety has been named a motivating factor for activism, which can lead to positive outcomes for youth (9). The combination of lack of agency and government inaction has led adolescents to participate in activism and other forms of civic engagement to combat climate changes, which may help to reduce eco-anxiety among youth (41).

A New Prescription: Youth Activism

Ballard and Ozer (41) define youth activism as the "organized efforts of young people to address the root causes of problems in their local, national, and global communities." Hickman's (40) global research implies that young people are frustrated with their elected officials for failing to protect their generation from the growing threat of climate change. With social media and digital platforms making activism easier and more accessible, youth are involved in many forms of activism and civic engagement, from smaller individual actions, e.g., signing petitions, changes in consumer behaviors, to larger organizational actions, e.g., lobbying, community organizing, conservation projects, lobbying for sustainable changes at schools. While these actions have many benefits for youth, they also have some potential drawbacks that should be noted.

Benefits of Activism for Youths

Beyond the benefits for the community and society that result from youth activism, there are also benefits for the adolescents themselves. Because sensation-seeking, or the desire to seek out novel experiences, increases in early adolescence and peaks at around 19 years of age, adolescents are likely to participate in either positive or negative risk-taking behaviors (30,41). Research suggests that engaging in positive risk-taking behaviors that require planning and impulse control can help improve self-regulation skills (32), personal responsibility, goal-setting, and social competence (42). Positive risk-taking may also protect against the negative emotional impacts of the climate crisis on youth. A literature review looking at positive risk-taking in adolescents found that it may help those suffering from mental health issues by reducing the internalization of symptoms, such as negative mood and low energy, developing resilience, and increasing positive affect, happiness, and optimism (42). Climate activism offers a positive, structured opportunity to meet youth's desire for novel, exciting activities while also supporting their development and well-being.

Activism itself can improve specific mental health outcomes and contribute to overall youth well-being. One study found that activism among adults who felt angry or frustrated about climate change led to improved mental health outcomes (5), and a second study found that participating in community engagement as a form of activism during adolescence and early young adulthood was linked to lower depressive symptoms in later adulthood (43). Organized action also can be a method for processing experiences related to the climate, connecting with others to cope with feelings of isolation, and managing anxiety (44). Eco-anxiety, in particular, can lead to a constructive response, highlighting the benefits of engaging youth in positive situations that reduce worry (45). Emerging literature has found that after climate-related disasters, children may experience post-traumatic growth if properly supported by caregivers to reflect, grieve, and articulate feelings on climate change (34). Furthermore, youth engagement in climate change activism can result in the growth of resilience, leadership skills, and collaboration (28). Overall positive well-being is also associated with greater pro-environmental actions, which may be due to the sense of autonomy and empowerment that engaging with activism can provide (5). See Figure 2 for a visualization of the relationship between adolescent eco-anxiety, climate activism, and experiences with the climate crisis.

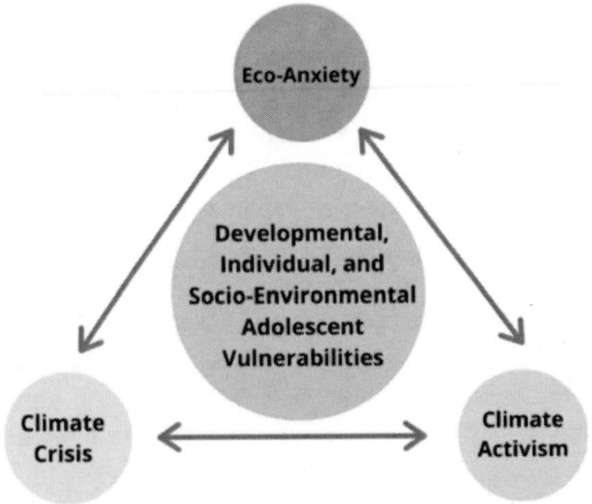

Figure 2. Factors influencing adolescent eco-anxiety and climate activism.

Empowerment has multiple dimensions, including cognitive empowerment, or the knowledge, skills, abilities, and resources needed to be active. Activism contributes to cognitive empowerment by giving youth an opportunity to learn about political processes and how to navigate these processes to achieve their goals (46). After completing a ten-session intervention about air quality issues and civic leadership, high school students in California reported increased self-efficacy related to using environmental data to improve their health, planning to address community issues, and taking action to address issues at the local and state level (47). Similar increases in sociopolitical self-efficacy were found following a Colorado high school-based intervention on youth participation in civic action (48).

A person's confidence in being able to make a positive change or contribution to the world around them, or emotional empowerment, can also be increased through activism (46). Activism is a way for youth to engage with social-political issues and increase their sense of autonomy despite their reliance on adults (5, 49). Acting in different roles within activism spaces also can contribute to identity formation for youth via positive risk-taking by facilitating the development of skills and confidence that can support them as adults. (42). Taking informed action can help adolescents feel part of a larger collective shift towards sustainability while also increasing personal motivation, self-confidence, and self-efficacy (50). Additionally, the opportunities for choice, autonomy, and self-direction in youth climate activism can increase the intrinsic motivation for climate change action, which can be further supported by providing positive performance feedback (11).

An additional dimension of empowerment, relational empowerment, refers to the belief in a community of people to make systemic change (46). Engaging with activism is a way for youth to contribute and connect to their community (41, 44). Participating in organized actions can help build stronger collective efficacy and foster a sense of community (51). Studies suggest that early adolescence is when civic involvement is developed (52), so this window offers a critical opportunity to engage adolescents' understanding and interest in climate change issues. The relationship between activism and empowerment, especially emotional and relational empowerment, can be affected by ethnic identity, especially for youth of color (53). Ethnic identity, which refers to a feeling of belonging and connection to an individual's ethnic group, can help increase youth empowerment and connections to their community, which in turn improves well-being (53). Activism also has demonstrated a mediating effect on the relationship between minority stress and health for youth of color and transgender, nonbinary (NB), and gender nonconforming (GNC) youth (49).

Drawbacks of Youth Activism

Although there is evidence that participating in activism can support youth well-being and improve mental health, there also is evidence that the sustained emotional and mental efforts required by activism can have a negative impact on mental health. Adult activists who live with HIV have reported greater depressive symptoms than non-activists living with HIV, but also reported greater well-being (54). In an analysis of longitudinal data, researchers found that political behaviors during adolescence and early young adulthood were weakly associated with increased depression in young adulthood (43). Similarly, the physical well-being of youth can also be negatively impacted by engagement in activism because of how stressful activism can be. Stress can be caused by youth activists' increased exposure to criticism and discrimination and their experiences of increased anger and burnout (5). Burnout is more likely when youth activists do not feel as though they are moving toward their goals or do not feel connected to the environmental movement (44).

The negative effects of activism can affect youth of color more than their counterparts. This is due in part to youth of color experiencing greater impacts of climate change, so the outcomes of activism, beneficial or not, will impact them more directly. They also face additional stress as a result of existing inequities and reduced access to activism opportunities, which can lead to burnout for youth of color more rapidly (5, 53). Indigenous youth, in particular, can face these stressors more often and more intensely due to how closely their cultural and personal identities are tied to the Earth (44).

Accounting for these impacts and working to reduce them is necessary to support sustainable youth activism. Youth activists cite the importance of adult support to reduce burnout and help achieve goals (44). This support could be mentoring, facilitating action, or providing connections to sources of power (44, 46). Adults also could support creating more systematic opportunities for activism to reduce the burden on individual youth activists to seek out opportunities or organize them if none are available. A high school-based sociopolitical development program facilitated by adult staff demonstrated positive outcomes, including increased efficacy and engagement, and was accessible to youth from marginalized backgrounds who had reduced access to similar programs (48). Another avenue is activism opportunities facilitated by faith-based organizations. Given that a variety of religions incorporate principles of social justice and/or respect for the natural world, faith-based organizations can support youth in both processing experiences of and feelings related to climate change and accessing activism (36). These organizations are often embedded in communities, so they understand community needs and are accessible to youth (36).

Both adults and youth can help prevent burnout by setting small measurable goals, celebrating achievements and efforts, and having regular check-ins with youth activists (44, 47). Check-ins are one way to build connections between activists and prioritizing this approach and other connection-centered activities can maximize the positive impacts of activism for young people. Adults, including activists, educators, and family members, can utilize their understanding of the motivations of youth activists to help reduce burnout and other pitfalls of activism by ensuring that outcomes align with youths' goals.

A Model for Motivating Youth Activism

Classic behavioral science models have been adapted to predict environmental behaviors. For example, Mancha and Yoder (55) adapted Azjen's (56) Theory of Planned Behavior into the Environmental Theory of Planned Behavior. A model that may be relevant to understanding youth motivations is the Integrative Social Identity Model of Collective Action, particularly in light of the importance of identity development and social relationships for adolescents.

The Integrative Social Identity Model of Collective Action

While forming their sense of identity, youth also are navigating how they relate to their peers, community, and society at large (57). The Integrative Social Identity Model of Collective Action (SIMCA) model explores how these interpersonal relationships and community identities affect youth motivation for activism by integrating perceptions of injustice, social identity, and collective efficacy (see Figure 3) (58). The SIMCA's use of social identity can help us understand the direct influence social identity has on the activism efforts of marginalized communities and their attitudes toward specific situations or social structures (58). Importantly, the integrative SIMCA model is more inclusive and specific than other models of collective action.

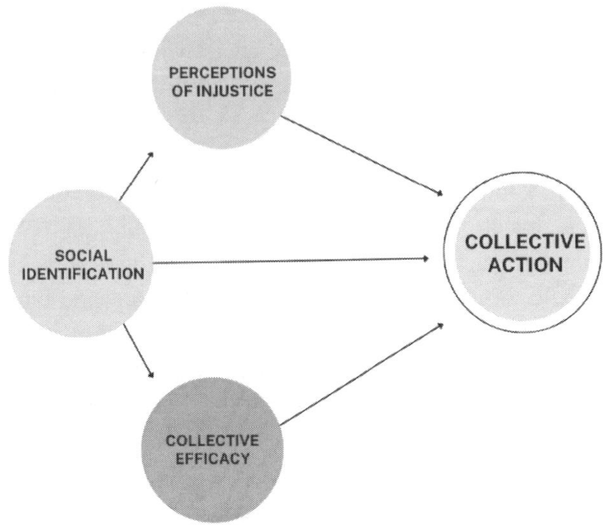

Figure 3. The Social Identity Model of Collective Action (58).

All three factors of SIMCA can be influenced by a person's sense of community (SOC), which is their sense of belonging and belief that community members will support one another (53). SOC can deepen a group's efficacy by giving them a sense of power and belief that they are capable of creating change within their community. Belonging to a community allows for opportunities to see activism modeled by others, which supports personal engagement and identification with activism. In their combined sample of German youth at large and youth at an environmental protest, Wallis and Loy (51) found correlations between the frequency of protesting and the perceived activism of their friends and their parents, for example.

Additionally, social identity can help individuals understand the need for collective action by focusing on their politicized identity, or problems people perceive in society and their attachments to those problems, which can have an effect on their perceived obligation to participate in social movements (59). Earnshaw et al. (54) investigated the relationship between stigma and activism among people living with HIV and found that those who reported experiencing stigma were more likely to identify as an activist compared to those who did not experience stigma. The implications of these findings highlight how past experiences of discrimination can be a motivating factor for activism. Although this study focused on adults, Earnshaw et al. (54) findings were supported by additional studies among Black adolescents and emerging adults.

Frequent experiences of racist events, racial microaggressions, and the resulting stress from these experiences among Black college students have been associated with greater political and civic engagement (60). Experiences of different types of racism also have been found to predict whether youth engage with high- or low-risk activism. Cultural racial discrimination among Black adolescents and emerging adults is related to a greater orientation toward low-risk activism, while experiences of institutional racial discrimination ignite high-risk activism for the same population (61). In Black adolescent boys who highly valued their Black identity as a part of their full identity, there was a relationship between the anticipation of racism and high-risk activism (60). This relationship is supported by the idea of minority stress, or the additional stress on minority groups as a result of socioeconomic inequities (49). A significant association between minority stress and activism also was found for youth of color, but not White youth (49).

Overall, the SIMCA model is a useful framework for understanding the role of social connectedness and encouraging activism among youth. However, existing behavioral models examine environmental behaviors in adults, but corresponding research on youth is limited. More research using models such as the ETPB and SIMCA is needed to foster and predict environmental behaviors among youth as youth activism is a powerful tool that may help shift policy and action on climate issues. For example, Greta Thunberg addressed the United Nations during the world's Climate Action Summit to demand answers about the failure to recognize the climate crisis, which resulted in motivating more than 60 countries to eliminate their carbon footprints by 2050 (62).

Youth Activism as a Tool to Combat Eco-Anxiety

The recent increase in youth movements and global calls to action to address the climate crisis make it critical that we have a better understanding of how to support and encourage youth. Throughout our literature search, we found that research on climate change activism among youth is very limited, with most research focused on youth from developed countries. A few studies have focused on encouraging youth to foster political interest and take action in climate justice efforts. Prime examples of activism-specific approaches are youth-led organizations and programs that aim to increase environmental health literacy through a focus on community-led activism (47, 63). Additional studies have focused more on social movements. Raducu and colleagues (64) highlighted how social campaigns are an essential component to increasing climate change discussions at a more local level to influence and create collective responses. Eide and Kunelius (65) synthesized interviews of youth activists from a global sample and found most youth activists viewed themselves as members of a global movement to combat climate change or as individuals who are fighting against the effects of climate change (e.g. due to geographic location) and are then considered activists due to their work. For instance, youth who speak at UN conferences about the global impact of climate change versus youth who are trying to combat climate change effects impacting local biodiversity are both activists, even if their initial reasons to take action were different (65). Overall, the existing literature on climate change activism among youth is sparse but slowly increasing, especially due to recent social media-driven opportunities for activism.

Social Media

With dramatic increases in global access, social media has become a key tool for activism among youth. The broad reach of social media platforms is being leveraged by youth to obtain knowledge, spread awareness, and rally support for climate change solutions and actions on a global scale. Adolescents are frequent users of digital media, with use increasing even more since the COVID-19 pandemic. In a 2021 poll by Commonsense Media, youth 13-18 years old reported spending an hour and a half on average per day on social media, with the most commonly ever-used social media sites being YouTube (83%), Instagram (70%), and TikTok (68%) (66). A content analysis of 100 English language videos sampled on TikTok tagged with #climatechange collectively received more than 200 million views, 40 million likes, and 600,000 comments (67). These data highlight the ease of disseminating information and reach through social media and the importance of ensuring accurate and reliable information on social media platforms. Not only is social media a way for youth to learn about environmental issues, but it also can be an avenue for youth activism. Recognizing how the power and promise of social media can be utilized to influence youth action against climate change can help health professionals and other adults to better leverage social media in their efforts to motivate and support youth to engage in climate activism.

Social Connection in Social Media

Peer relationships take on enormous significance during adolescence, and adolescents are particularly prone to turn to their peers for support and social modeling (68). Adolescent social networks are heavily influenced by social media, providing connection and mediation of content consumption based on what is trending and what their peers are sharing. Recently, more attention has been paid in the literature to using a social connectedness approach that focuses on social networks to try to understand civic engagement among adolescents.

Haslam et al. (69) define social connectedness as the sense of belonging and subjective psychological bond that people feel towards individuals and groups of people. Social connectedness and social networks are closely related, and positive behaviors among some social network members encourage positive health norms among all members. Social Capital Theory (70) explains that youth with multiple peer connections have more social resources that can provide opportunities to participate in civic action and increase feelings of social responsibility (71). Conversely, low social capital exists when there is a lack of networks, rules, and roles, and also could stem from a lack of trust and norms that promote low collaboration and antisocial behavior (72). Social media can be used to raise awareness of group-based efforts, minimizing the risk that an individual feels alone in striving for change and helping to reduce burnout among activists. Additionally, youth activist networks on social media can help combat overwhelming feelings of helplessness and reduce the perception of limited action by peers (73), which may contribute to an increase in youth activism.

Cyberactivism has become increasingly popular and highlights the shift in our society from past forms of grassroots activism to passive social media activism such as liking and sharing content. These passive forms of activism may seem to increase awareness among adolescents, but adolescents' involvement in grassroots activism may decline due to increased feelings of fulfillment in their passive activism (74). Additionally, engagement with movements on social media sites may not promote offline activist behavior (75). There is concern that there is a disconnect between awareness of social movements and actions to enact change that may have an impact on offline behavior. This creates a challenge when the easiest way to reach a large audience and motivate youth is through social media. Health professionals and organizations who are at the forefront of the climate crisis should take more varied approaches beyond social media to motivate and engage youth to participate in more hands-on forms of activism.

Education and Communication for Youths to Inspire Climate Change Activism

It may be valuable to leverage this unique period of cognitive and social development to stimulate inspiration and climate action among youth. Due to an increase in neural interconnectedness that allows for more complex and sophisticated thinking (76), along with increased opportunities for freedom, decision-making and actions become more independent during adolescence. During this time, adolescents develop their sense of self-identity, establishing personal values around their commitments, goals, motivations, and psychosocial well-being (76). Therefore, the adolescent period is a critical time for intervention. An adolescent's socio-political ideology may still be fluid and is more likely to overcome socio-ideological biases and barriers that contribute to polarization over climate action found in adults (77). Given adolescents' orientation to peers, peer-led approaches like peer education have been successful in adolescent sexual health and substance use programs, and they also may be working to encourage youth climate activism (78). Kerrigan and Weiss (79) proposed that peer education can effect change at the group or societal level by modifying norms and incorporating aspects of collective action that contribute to individual change. Importantly, research suggests that promoting youth action is a complex educational process that must be embedded in a more meaningful social context, rather than a simple teacher-learner interaction (80). Researchers argue that empowering youth to address climate change is more important than understanding climate change, as students not only need the facts but also the resources and motivation to take action (81, 82).

A synthesis of recent literature reviews on climate change education (CCE) highlights key considerations to increase youth engagement, with a unifying theme of the need for educators to go beyond just relaying facts on climate change (81-83). These reviews mentioned the importance of experiential learning and fostering a connection to the subject through personally relevant examples. The inclusion of culturally relevant content was cited as particularly beneficial for historically marginalized, at-risk, or under-served students (83). An interesting emerging area of research is the relevance of knowledge from Indigenous societies for CCE (81), which offers both a different context to view this complex topic and awareness of environmental justice. Supporting learner autonomy and agency proved valuable to deepen understanding, including having the opportunity to experience the scientific process for themselves and engaging in designing or implementing school or community projects related to climate change (82). Collaborations with scientists served to not only engage students but also increased the confidence of educators in facilitating student exploration and empowered students to learn more and take action (82). All three reviews also addressed in some way the socio-emotional context of this subject: the importance of positive framing (83), the need for affect-driven approaches to shift attitudes and actions (81), and the value of deliberative discussion to facilitate understanding other viewpoints and recognition of knowledge gaps, potentially opening the door to address misinformation/misconceptions (82). These reviews also highlight the growing tension between knowledge-based CCE programs and the more complex but potentially more effective interdisciplinary and experiential approaches, particularly in primary and secondary education classrooms. Rousell and Cutter-Mackenzie-Knowles (81) argue that an interdisciplinary approach is necessary to adequately prepare students to understand and address the social, ethical, political, and scientific complexities of climate change.

The development of educational exhibitions or materials by students, whether for adults or peers, has been shown to contribute to transformational learning in students and provides educational gains for the observers. Transformational learning is defined as a perspective transformation, where learning leads to critically examining prior interpretations and assumptions to achieve a new meaning (50). This may be critical for reaching the broad acceptance of climate change necessary to meet the collective action needs of our current climate crisis. Studies that combine scientific knowledge and artistic mediums seem to be particularly effective at achieving transformative learning (84). Examples in the literature that highlight this effectiveness range from primary to tertiary education level students, and across artistic mediums and countries (11, 50, 84). Art has a long history of facilitating meaning-making and civic engagement, so the engagement of arts and humanities programs in the climate discourse may be an important lever to help shift the hearts and minds of society on this issue.

Actionable Steps

Reviewing the available literature has provided a better understanding of how climate change affects the mental health of youth, how activism can address these effects, and how this knowledge can be translated into action in the practice, research, and policy spheres. The practice-focused results of our review demonstrate the need for the recognition of eco-anxiety as a legitimate and serious mental health concern among youth as well as the need for improved, accessible environmental education and positive activism opportunities for youth. This scoping review also validated youth climate change activism as a growing field with multiple areas for future research. Policy changes that elevate the voices of young people are necessary to alleviate eco-anxiety and directly address climate change. Based on our findings, we recommend the following steps be taken to promote youth well-being:

Research Recommendations

1. Prioritize young people, underrepresented minorities, Indigenous youth, and youth from the Global South as participants in studies on climate change
2. Conduct rigorous research to investigate direct and indirect relationships between youth activism and feelings of eco-anxiety
3. Research potential programs that effectively motivate youth to climate change action
4. Investigate whether climate change activism functions as positive risk-taking behavior and/or an alternative to negative risk-taking behavior
5. Discern the efficacy of new behavioral models that explain climate change activism among youth

Practice Recommendations

1. Add Eco-Anxiety to the DSM-5R or DSM-6
2. Encourage adults at the organizational, community, and institutional levels to create systematic opportunities for ongoing youth activism

3. Incorporate interdisciplinary knowledge and practice into youth education on climate change
4. Add transformational approaches to climate change education for youth

Policy Recommendations

1. Involve youth in decision-making processes about climate change policies
2. Integrate climate change curricula across subjects and disciplines in educational settings
3. Return lands to Indigenous communities to provide environmental security and prioritize the health and well-being of Indigenous youth

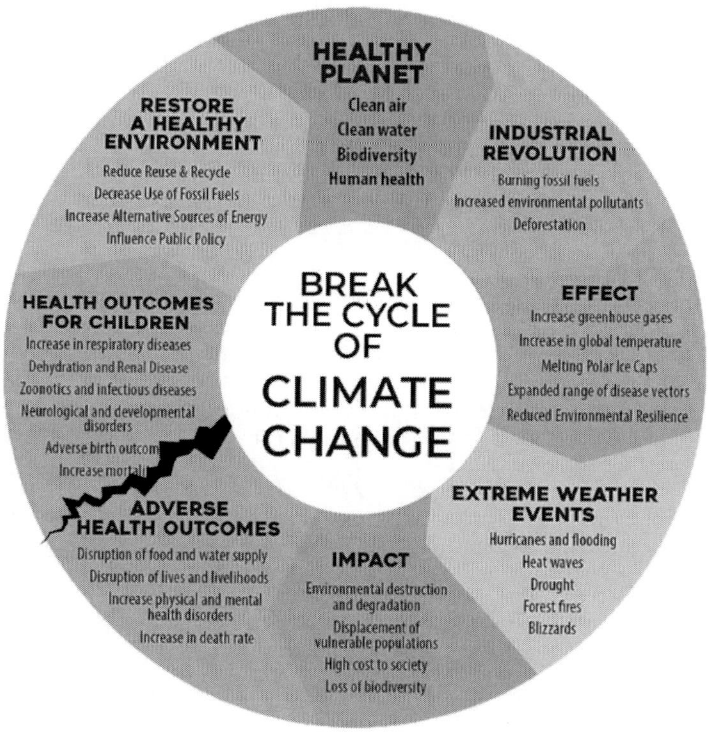

Figure 4. Cycle of environmental health disparities.

Strengths and Limitations

A major limitation of this review is that it is centered on areas of emerging research, which is why the scoping review format was chosen. Surprisingly, there were few articles found in the literature that focused on adolescents, eco-anxiety, and activism, although the research base is growing. The available literature on youth climate activism, underrepresented and disadvantaged youth, and literature on encouraging youth climate activism is limited. One of the major gaps in existing literature includes the lack of research on youth from developing countries and youth who have historically been underrepresented in existing spaces about climate change. It is possible that our literature search missed some of the available research, but we used a variety of different terms to capture a broad variety of studies. However, we limited our search to only articles in English, which may have restricted us to a narrower range of articles from non-English-speaking countries. A strength of this review was its cross-disciplinary nature and developmental lens to contextualize eco-anxiety, youth activism, and climate communication within a better understanding of adolescents' needs. Some of the research we located was in education or environmental science journals. This is an area of opportunity for public health professionals, so we can better understand the public health implications of youth interventions for climate change and activism.

Conclusion

Through this scoping review, we identified gaps in the current literature about eco-anxiety in adolescents, the benefits and drawbacks of climate activism, and methods to engage youth in climate activism. To our knowledge, this is the first paper that explores activism as a "treatment" option for reducing eco-anxiety among adolescents, which is based on our synthesis of the literature on encouraging youth activism and using appropriate communication methods to motivate youth action. Per Figure 4, our research addresses the effects of climate change on adverse health outcomes and provides solutions that are crucial to breaking the cycle of environmental health disparities. Future research should focus on eco-anxiety among young people and explore potential differences between youth and adults in risk, contributing factors, symptomatology, and outcomes. The results of this research can inform effective public health practices and policies that will benefit young people.

As we shift our focus to the societal impacts of climate change, it is critical that we elevate the issue of eco-anxiety and protect young people and the future. As we shift our focus to the societal impacts of climate change, it is critical that we elevate the issue of eco-anxiety and protect young people and the future.

Acknowledgments

This project was supported by the Break the Cycle of Children's Environmental Health Disparities Program of the Southeast PEHSU at Emory University and Break the Cycle of Health Disparities, Inc, as well as the University of San Francisco's School of Nursing and Health Professions. We also want to acknowledge all of the inspiring youth activists who are leading the fight against climate change. The authors have no conflicts to declare.

Appendix A: Literature Search Criteria

Search Engines	Ebsco Discovery Service (Fusion), APA PsychNet, PubMed, Google Scholar, USF Gleeson Library
Search Terms	Population • adolescents or teens or young adults or young people or youth • Youth activist or youth climate activist or climate activist • Black, minority and/or youth of color, Indigenous or Native or Aboriginal or Indians or First Nations • Hispanic and/or Latino Concept • adolescent health or well-being or youth health impacts • activism or action or community engagement • Climate change or climatic changes or greenhouse effect or global warming or weather • Climate anxiety or eco-anxiety or eco-depression or environmental distress • Environmental or environment Context • IPCC or IPCC report • Motivation • Benefits • content analysis, communication, extended parallel process model, action competence in sustainable development, transformative learning theory
Eligibility Criteria	Available in English, Published between 2000-2022, Accessible through the University of San Francisco, Relevant to research questions

Appendix B

Sample Search Strategy for Ebsco Discovery Service: Indigenous or Native or Aboriginal or Indians or First Nations AND Activism AND Youth.

References

[1] Intergovernmental Panel on Climate Change (IPCC). IPCC WGII Sixth Assessment Report, 2022. URL: https://www.ipcc.ch/report/ar6/wg2/.

[2] Sanson AV, Van Hoorn J, Burke SE. Responding to the impacts of the climate crisis on children and youth. Child Dev Perspect 2019;13(4):201-7.

[3] Intergovernmental Panel on Climate Change (IPCC). Summary for Policymakers. In: Global Warming of 1.5°C. An IPCC Special Report on the impacts of global warming of 1.5°C above pre-industrial levels and related global greenhouse gas emission pathways, in the context of strengthening the global response to the threat of climate change, sustainable development, and efforts to eradicate poverty, 2018. URL: doi:10.1017/9781009157940.001.

[4] Burke SE, Sanson AV, Van Hoorn J. The psychological effects of climate change on children. Curr Psychiatr Rep 2018 May;20(5):1-8.

[5] van Nieuwenhuizen A, Hudson K, Chen X, Hwong AR. The effects of climate change on child and adolescent mental health: clinical considerations. Curr Psychiatr Rep 2021;23(12):1-9.

[6] Ma T, Moore J, Cleary A. Climate change impacts on the mental health and wellbeing of young people: A scoping review of risk and protective factors. Soc Sci Med 2022;18:114888.

[7] To P, Eboreime E, Agyapong VI. The impact of wildfires on mental health: A scoping review. Behav Sci 2021;11(9):126.

[8] Cianconi P, Betrò S, Janiri L. The impact of climate change on mental health: a systematic descriptive review. Fron Psychiatry 2020;11:74.

[9] Coffey Y, Bhullar N, Durkin J, Islam MS, Usher K. Understanding eco-anxiety: A systematic scoping review of current literature and identified knowledge gaps. J Climate Change Health 2021;3:100047.

[10] Sabherwal A, Ballew MT, van Der Linden S, Gustafson A, Goldberg MH, Maibach EW, et al. The Greta Thunberg Effect: Familiarity with Greta Thunberg predicts intentions to engage in climate activism in the United States. J Appl Soc Psychol 2021;51(4):321-33.

[11] Kowasch M, Cruz JP, Reis P, Gericke N, Kicker K. Climate youth activism initiatives: Motivations and aims, and the potential to integrate climate activism into ESD and transformative learning. Sustainability 2021;13(21):11581.

[12] Gutierrez-Bucheli L, Reid A, Kidman G. Scoping reviews: Their development and application in environmental and sustainable education research. Environ Educ Res 2022;28(5):645-73.

[13] Albrecht G. Psychoterratic conditions in a scientific and technological world. Ecopsychol 2012:241-64.

[14] Pihkala P. Anxiety and the ecological crisis: An analysis of eco-anxiety and climate anxiety. Sustainability 2020;12(19):7836.

[15] Schreiber M. Addressing climate change concerns in practice. Am Psychol Assoc 2021. URL: https://www.apa.org/monitor/2021/03/ce-climate-change.

[16] Clayton S. Climate anxiety: Psychological responses to climate change. J Anxiety Disord 2020;74:102263.

[17] Clayton S, Karazsia BT. Development and validation of a measure of climate change anxiety. J Environ Psychol 2020;69:101434.

[18] Comtesse H, Ertl V, Hengst SM, Rosner R, Smid GE. Ecological grief as a response to environmental change: a mental health risk or functional response? Int J Environ Res Public Health 2021;18(2):734.

[19] Doherty TJ, Clayton S. The psychological impacts of global climate change. Am Psychol 2011;66(4):265.

[20] Gifford E, Gifford R. The largely unacknowledged impact of climate change on mental health. Bull Atomic Scientists 2016;72(5):292-7.

[21] Hickman C. We need to (find a way to) talk about Eco-anxiety. J Soc Work Pract 2020;34(4):411-24.

[22] Dockett L. Clinician's Digest: The rise of eco-anxiety'. Psychother Networker 2019;43(10):11-4.

[23] Pihkala P. Eco-amxiety, tragedy, and hope: Psychologican alnd spiritual dimensions of climate change. Zygon® 2018;53(2):545-69.

[24] Schwartz SE, Benoit L, Clayton S, Parnes MF, Swenson L, Lowe SR. Climate change anxiety and mental health: Environmental activism as buffer. Curr Psychol 2022;28:1-4.

[25] Stanley SK, Hogg TL, Leviston Z, Walker I. From anger to action: Differential impacts of eco-anxiety, eco-depression, and eco-anger on climate action and wellbeing. J Climate Change Health 2021;1:100003.

[26] Usher K, Durkin J, Bhullar N. Eco-anxiety: How thinking about climate change-related environmental decline is affecting our mental health. Int J Ment Health Nurs 2019;28(6):1233-4.

[27] Baudon P, Jachens L. A scoping review of interventions for the treatment of eco-anxiety. Int J Environ Res Public Health 2021;18(18):9636.

[28] Cox RS, Hill TT, Plush T, Heykoop C, Tremblay C. More than a checkbox: engaging youth in disaster risk reduction and resilience in Canada. Natural Hazards 2019;98(1):213-27.

[29] Naar S, Suarez M. Motivational interviewing with adolescents and young adults. New York: Guilford, 2021.

[30] Jensen F, Nutt AE. The teenage brain: A neuroscientist's survival guide to raising adolescents and young adults. New York: Harper, 2015.

[31] Lawler J, Patel M. Exploring children's vulnerability to climate change and their role in advancing climate change adaptation in East Asia and the Pacific. Environ Dev 2012;3:123-36.

[32] National Academies of Sciences, Engineering, and Medicine. Promoting positive adolescent health behaviors and outcomes: Thriving in the 21st century. Washington, DC: National Academies Press, 2020.

[33] La Greca AM, Lai BS, Joormann J, Auslander BB, Short MA. Children's risk and resilience following a natural disaster: Genetic vulnerability, posttraumatic stress, and depression. J Affect Disord 2013;151(3):860-7.

[34] Clayton S, Manning CM, Krygsman K, Speiser M. Mental health and our changing climate: Impacts, implications, and guidance. Washington, DC: American Psychological Association, 2017.

[35] Patnaik A, Son J, Feng A, Ade C. Racial disparities and climate change. Princeton, NJ: Princegton UniversityPress, 2020.

[36] Sachdeva S. Religious identity, beliefs, and views about climate change. Oxford Research Encyclopedia of Climate Science, 2016. doi. org/10.1093/acrefore/ 9780190228620.013.

[37] Helldén D, Andersson C, Nilsson M, Ebi KL, Friberg P, Alfvén T. Climate change and child health: a scoping review and an expanded conceptual framework. Lancet Planetary Health 2021;5(3):e164-75.

[38] Chalupka S, Anderko L, Pennea E. Climate change, climate justice, and children's mental health: A generation at risk? Environ Justice 2020;13(1):10-4.

[39] Ragavan MI, Marcil LE, Garg A. Climate change as a social determinant of health. Pediatrics 2020;145(5):e20193169.

[40] Hickman C, Marks E, Pihkala P, Clayton S, Lewandowski ER, Mayall EE, et al. Young people's voices on climate anxiety, government betrayal and moral injury: A global phenomenon. Government Betrayal and Moral Injury: A Global Phenomenon, 2021. http://dx.doi.org/10.2139/ssrn.3918955.

[41] Ballard PJ, Ozer EJ. The implications of youth activism for health and well-being. In: Conner J, Rosen SM, eds. Contemporary youth activism: Advancing social justice in the United States. Santa Barbara, CA: ABC-CLIO, 2016:223-44.

[42] Duell N, Steinberg L. Adolescents take positive risks, too. Dev Rev 2021;62:100984.

[43] Wray-Lake L, Shuber J, Starr LL, Starr LR. Examining associations between civic engagement and depressive symptoms from adolescence to young adulthood in a national US sample. Appl Dev Sci 2017;23(2):119-31.

[44] Godden NJ, Farrant BM, Yallup Farrant J, Heyink E, Carot Collins E, Burgemeister B, et al. Climate change, activism, and supporting the mental health of children and young people: Perspectives from Western Australia. J Paediatr Child Health 2021;57(11):1759-64.

[45] Verplanken B, Marks E, Dobromir AI. On the nature of eco-anxiety: How constructive or unconstructive is habitual worry about global warming? J Environ Psychol 2020;72:101528.

[46] Wray-Lake L, Abrams LS. Pathways to civic engagement among urban youth of color. Monogr Soc Res Child 2020;85(2):7-154.

[47] Madrigal D, Claustro M, Wong M, Bejarano E, Olmedo L, English P. Developing youth environmental health literacy and civic leadership through community air

monitoring in Imperial County, California. Int J Environ Res Public Health 2020;17(5):1537.

[48] Hipolito-Delgado CP, Stickney D, Zion S, Kirshner B. Transformative student voice for sociopolitical development: Developing youth of color as political actors. J Res Adolesc 2022;32(3):1098-108.

[49] Frost DM, Fine M, Torre ME, Cabana A. Minority stress, activism, and health in the context of economic precarity: Results from a national participatory action survey of lesbian, gay, bisexual, transgender, queer, and gender non-conforming youth. Am J Commun Psychol 2019;63(3-4):511-26.

[50] Trott CD. Climate change education for transformation: Exploring the affective and attitudinal dimensions of children's learning and action. Environ Educ Res 2022;28(7):1023-42.

[51] Wallis H, Loy LS. What drives pro-environmental activism of young people? A survey study on the Fridays For Future movement. J Environ Psychol 2021;74:101581.

[52] Smetana JG, Campione-Barr N, Metzger A. Adolescent development in interpersonal and societal contexts. Annu Rev Psychol 2006;57:255.

[53] Lardier Jr DT. An examination of ethnic identity as a mediator of the effects of community participation and neighborhood sense of community on psychological empowerment among urban youth of color. J Commun Psychol 2018;46(5):551-66.

[54] Earnshaw VA, Rosenthal L, Lang SM. Stigma, activism, and well-being among people living with HIV. AIDS Care 2016;28(6):717-21.

[55] Mancha RM, Yoder CY. Cultural antecedents of green behavioral intent: An environmental theory of planned behavior. J Environ Psychol 2015;43:145–54.

[56] Ajzen I. From intentions to actions: A theory of planned behavior. In: Kuhl J, Beckmann J, eds. Action control. Berlin: Springer, 1985:11–39.

[57] Flanagan C, Levine P. Civic engagement and the transition to adulthood. Future Child 2010;20(1):159-79.

[58] van Zomeren M, Postmes T, Spears R. Toward an integrative social identity model of collective action: a quantitative research synthesis of three socio-psychological perspectives. Psychol Bull 2008;134(4):504.

[59] Sturmer S, Simon B. Collective action: Towards a dual-pathway model. Eur Rev Soc Psychol 2004;15(1):59-99.

[60] Hope EC, Cryer-Coupet QR, Stokes MN. Race-related stress, racial identity, and activism among young Black men: A person-centered approach. Dev Psychol 2020;56(8):1484.

[61] Hope EC, Gugwor R, Riddick KN, Pender KN. Engaged against the machine: Institutional and cultural racial discrimination and racial identity as predictors of activism orientation among Black youth. Am J Commun Psychol 2019;63(1-2): 61-72.

[62] Gonzalez B. Youth activism is creating lasting change. USC Annenberg School For Communication and Journalism, 2020. URL: https://annenberg.usc.edu/research/center-public-relations/usc-annenberg-relevance-report/youth-activism-creating-lasting.

[63] Hilder C, Collin P. The role of youth-led activist organisations for contemporary climate activism: the case of the Australian Youth Climate Coalition. J Youth Stud 2022;25(6): 793-811.
[64] Raducu R, Soare C, Chichirez CM, Purcarea MR. Climate change and social campaigns. J Med Life 2020;13(4):454.
[65] Eide E, Kunelius R. Voices of a generation the communicative power of youth activism. Climatic Change 2021;169(1):1-20.
[66] Rideout V, Peebles A, Mann S, Robb M. The common sense census: Media use by tweens and teens. San Francisco, CA: Common Sense, 2022.
[67] Basch CH, Yalamanchili B, Fera J. Climate change on TikTok: A content analysis of videos. J Commun Health 2022;47(1):163-7.
[68] Kilford EJ, Garrett E, Blakemore SJ. The development of social cognition in adolescence: An integrated perspective. Neurosci Biobehav Rev 2016;70:106-20.
[69] Haslam C, Cruwys T, Haslam SA, Jetten J. Social connectedness and health. Encyclopaedia ZGeropsychology 2015;2015:46-1.
[70] Bourdieu P. The forms of capital. In: Richardson JG, ed. Handbook of theory and research for the Sociology of Education. Westport, CT: Greenwood, 1986;241-58.
[71] Oosterhoff B, Alvis L, Deutchman D, Poppler A, Palmer CA. Civic development within the peer context: Associations between early adolescent social connectedness and civic engagement. J Youth Adolesc 2021;50(9):1870-83.
[72] Claridge T. Exploring the limits of social capital: Can social capital be continually improved or is there a maximum? Dunedin, NZ: Institute of Social Capital, 2022.
[73] Fritsche I, Masson T. Collective climate action: When do people turn into collective environmental agents? Curr Opin Psychol 2021;42:114-9.
[74] Yankah S, Adams KS, Grimes L, Price A. Age and online social media behavior in prediction of social activism orientation. J Soc Media Soc 2017;6(2):56-89.
[75] Kristofferson K, White K, Peloza J. The nature of slacktivism: How the social observability of an initial act of token support affects subsequent prosocial action. J Consumer Res 2014;40(6):1149-66.
[76] Adolescent development. URL: https://youth.gov/youth-topics/adolescent-health/adolescent-development.
[77] Lawson DF, Stevenson KT, Peterson MN, Carrier SJ, L Strnad R, Seekamp E. Children can foster climate change concern among their parents. Nature Climate Change 2019;9(6):458-62.
[78] Abdi F, Simbar M. The peer education approach in adolescents-narrative review article. Iranian J Public Health 2013;42(11):1200.
[79] Kerrigan D, Weiss E. Peer education and HIV/AIDS: Past experience, future directions. Washington, DC: Population Council, 2000
[80] De Vreede C, Warner A, Pitter R. Facilitating youth to take sustainability actions: The potential of peer education. J Environ Educ 2014;45(1):37-56.
[81] Rousell D, Cutter-Mackenzie-Knowles A. A systematic review of climate change education: Giving children and young people a 'voice'and a 'hand'in redressing climate change. Child Geographies 2020;18(2):191-208.

[82] Monroe MC, Plate RR, Oxarart A, Bowers A, Chaves WA. Identifying effective climate change education strategies: A systematic review of the research. Environ Educ Res 2019;25(6):791-812.

[83] Merritt EG, Stern MJ, Powell RB, Frensley BT. A systematic literature review to identify evidence-based principles to improve online environmental education. Environ Educ Res 2022;28(5):674-94.

[84] Bentz J. Learning about climate change in, with and through art. Climatic Change 2020;162(3):1595-612.

Chapter 3

Acting on Power and Health Inequities: An Essential Approach to Global Environmental Challenges

Carlos A. Faerron Guzmán[1,2,3,*], MD, MSc and Olena Zotova[4], MDCM

[1]Graduate School, University of Maryland, Baltimore, MD, United States of America
[2]Planetary Health Alliance, Harvard TH Chan School of Public Health, Boston, Massachusetts, United States of America, [3]ISGlobal, Barcelona, Spain
[4]Pompeu Fabra University, Spain

Abstract

Human civilization has less than ten years remaining to limit environmental changes that risk undoing a century of hard-earned gains in global health and human wellbeing. Current actions are insufficient to address the health emergency caused by global environmental change. We provide a diagrammatic intersectional analysis of the interaction between power, health, and environmental change that explains the prevailing status quo in environmental action, despite the health losses experienced by "lower power" groups. Using real-world examples, we demonstrate how the power-health-environment nexus underlies current delays in redressing health and environmental inequities and how we must act on this nexus to protect global and planetary health from anthropogenic disruptions in Earth's natural systems.

---

* ***Correspondence:*** Carlos A. Faerron Guzmán, 621 W Lombard St, Suite 334, Baltimore, MD 21201. The University of Maryland, Baltimore, Graduate School.
Email: cfaerronguzman@umaryland.edu.

In: Environmental Health: Leaders for Climate Change
Editors: I Leslie Rubin and Joav Merrick
ISBN: 979-8-89530-347-4

Introduction

Remarkable improvements in human health and wellbeing have been achieved in parts of the global population over the last two centuries. In this sliver of human history, our species has managed to drastically reduce the number of people affected by perils that once shaped communities all over the world (1). Global development efforts have led to drastic declines in child mortality and maternal mortality. Infectious diseases, once the biggest killer globally, have been significantly reduced with widespread access to clean water and sanitation. Antibiotics, vaccines, oral rehydration solutions, and many other medical advances have saved millions of lives just in the last seven decades. Life expectancy on all continents has been steadily rising (2). Technology, particularly mechanized farming and synthetic fertilizers, as well as land reform, have enhanced the performance of our food systems to support global nutrition needs. Undernourishment has steadily declined despite population growth (3). The reach of education and access to knowledge are also on the rise. Universal primary education is a close reality in almost every country in the world, and there is progress in closing the gender gap in education at all levels (4).

However, global progress, coupled with a growing population and increased consumption of natural resources, has also disrupted Earth's natural systems. Climate change, ocean acidification, soil degradation, forest loss, water pollution, and scarcity, as well as biodiversity loss, are examples of such disruptions (5). Paradoxically, as natural environments degrade, human wellbeing continues to improve in a proportion of the global population, a phenomenon which has been termed the "environmentalist's paradox" (6). Meanwhile, a growing proportion of the global population is negatively affected by the decline of Earth's natural systems. What global averages fail to show are the rapid, uneven, and unjust health and wellbeing losses suffered by some population groups. An increasingly large number of people are negatively affected by escalating degrees of environmental changes, and this number is likely to increase. We argue that an intersectional power imbalance (see box 1) drives these inequitable consequences of environmental changes and exacerbates health inequities globally, as it simultaneously delays action toward a more sustainable present and future (see Figure 1).

The status quo scenario implies insufficient action to impede a continuous degradation of Earth's natural systems and minor or no changes to current rates of natural resource extraction and technological advancement, as well as to the underlying consumption and demographic trends.

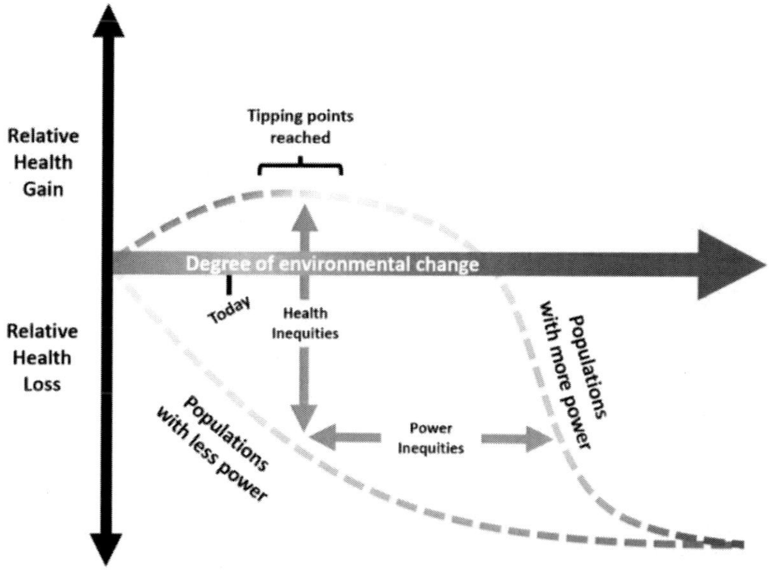

Figure 1. Current and future trends of power and health inequities in a status quo scenario of unabated global environmental change. Health Impacts of Anthropogenic Environmental Change – Status quo scenario. Inequitable health impacts and inequitable power distribution explain the delay to act and inaction from the most powerful in the face of global environmental change.

Figure 1 is based on the following six observations:

1. Populations with less power tend to be more affected by environmental change *(figure 1; lower line, health inequities arrow)*
2. Those who are more affected by environmental change have less power to act on it *(figure 1; lower line, power inequities arrow)*
3. Lower power groups can suffer disproportionate adverse consequences from actions addressing global environmental changes *(figure 1; lower line dropping sharply into "relative health loss")*
4. Issues disproportionately affecting lower power groups are insufficiently addressed by the global community *(figure 1; lower line within "relative health loss." Lack of action is inferred in the progression of the "degree of environmental change" arrow to the right.)*
5. Higher power groups are delaying action to address the health emergency caused by global environmental change *(figure1; lack of*

action is inferred in the progression of the "degree of environmental change" arrow to the righ.)

6. As the degree of environmental change increases, everyone including higher power groups will suffer adverse health impacts *(figure 1; "tipping points reached," lower and upper lines dropping into "relative health loss" along the "degree of environmental change" arrow)*

Each one of these observations is hereafter exemplified through a diversity of salient cases of power, health, and environmental inequities at global and local levels. The examples were chosen for their saliency and relevance in the current global context while noting a few limitations. First, although we compiled a diverse set of cases which will demonstrate the broad applicability of Figure 1, we recognize that the examples cited henceforth do not exhaustively represent the full scope of interactions between power, environment, and health around the world. Moreover, while we argue for a multidimensional power analysis (see box 1), we note that current literature in this realm remains scarce. As a result, most of our case examples consider only a unidimensional power analysis, whereby we use commonly cited individual and population-level characteristics to identify groups of relatively higher or lower power. By exemplifying a broad range of interlinked complex circumstances that can add or remove power to/from people, we hope to inspire a more comprehensive consideration of power analysis in future efforts in the environmental justice space. Overall, Figure 1 depicts how a growing degree of environmental change is interdependent with growing power and health inequities in a status-quo scenario.

Box 1. Power: An essential concept to understand the Anthropocene

At the core of Figure 1 lies the understanding of power imbalances. Power, for the sake of this analysis, is a multidimensional concept that refers to the individual or collective ability to produce a desired outcome, i.e., to be able to (power, from the Latin word, potere: "to be able") (7).

"Power" encompasses many concepts, and it enables us to perform a more comprehensive analysis of environmental and health issues as compared to limiting ourselves to more specific unidimensional concepts/categories. Categorizing people using unidimensional characteristics can oversimplify complex interactions, intersections, and gradients of determinants of health across populations. Nonetheless, most analyses of environmentally driven health/wellbeing inequities still rely on more specific characteristics pertaining to, among other determinants:

- income (e.g., people living in low-income/high-income regions/nations; people living in poverty/people not living in poverty)
- met needs (e.g., people living with unmet needs/people with needs fulfilled)
- participation and representation (e.g., people that are marginalized/included);
- race/ethnicity/gender/sexuality/religion (e.g., minorities/hegemonic populations; people of color/European descent; heteronormative/LGBTQIA2S+)
- age (e.g., unborn/children/adults/elders)
- related to resilience/vulnerability (e.g., people who are vulnerable/resilient)
- status of development (e.g., people living in developing/developed regions/nations; low development/high development regions/nations; industrializing/industrialized regions/nations)
- geographic or historic variables (e.g., global north/global south; the west/the rest; first world/third world)
- social capital (e.g., people with low social capital/high social capital)
- access to technology (people with low access to technology/high access to technology); among others

We use the aforementioned as the most salient examples of unidimensional power analyses that are found in the literature while noting that many of these represent outdated and sometimes racist, colorist, ableist, sexist, and aporophobic perspectives.
As such, we argue for a dynamic intersectional power analysis that considers all these categories/concepts. We consider power to be a product of all the former variables (e.g., race, ethnicity, socioeconomic status, gender, religion, nationality, sexual orientation, etc.) in consideration of time and space. Indeed, power depends not only on the characteristics mentioned above but also changes as people navigate different spaces and different moments (i.e., who am I? where am I? when am I?). We consider that a unidimensional analysis of power, as is prevalent in the literature (see below for examples), and how it influences health outcomes in the face of global environmental challenges masks real-world differences and nuances between populations within nation-states and regions. We note that power is closely related to the concept of agency, which can be understood as "the capacity of an individual to actively and independently choose and to affect change; free will or self-determination"(8). Nonetheless, in our analysis, we use the concept of power referring not only to individuals but also to collectives. Lastly and perhaps most importantly, when using terminology such as "populations with less power" or "lower power groups", we do not mean to characterize these populations as helpless, passive, or incapable. On the contrary, we recognize the immense courage and resilience with which these populations are attempting to improve their lives while facing global environmental changes. We also acknowledge that people in "lower power groups" are not there because of innate characteristics or faults of their own but as a product of the structural makeup of the societies they live in.

Observation 1: Populations with Less Power Tend to Be More Affected by Environmental Change

At ten years old, Hibo had to leave her home in search of food and water as an unprecedented drought exacerbated by climate change ravaged her country, Somalia. Around the world, billions of children like Hibo are extremely vulnerable to worsening environmental threats (9, 10).

Figure 2. Somali region drought response © UNICEF Ethiopia/2022/Mulugeta Ayene, CC BY-NC-ND 2.0 (11).

Evidence clearly shows that the impacts of global environmental change are uneven and unjust. While population groups with more power largely benefit from resource over-extraction driven by industrial and technological development and globalization, they often do so at the expense of the health and wellbeing of people with less power (12). As a consequence, populations with less power are disproportionately bearing the brunt of the impacts of global environmental change. In that regard, the so-called "environmentalist's paradox" described in the literature (6) arises from a privileged perspective on global environmental change, as it seems to only apply to the populations with more power. There is no environmentalist paradox for populations with less power, who already suffer health losses due to global environmental degradation. This unequal burden can be attributed to three overlapping adverse conditions impacting lower power populations: 1) higher exposure to environmental threats and disturbances, 2) higher vulnerability, and 3) lower resilience to global environmental change (13–16). Multiple reasons underlie these adverse conditions and must be explored through an intersectional power analysis and a lens of eco-social determinants of health. How people face high temperatures, sea-level rise, droughts, floodings, loss of biodiversity, and pollution, to name a few environmental disturbances, is determined by their income, their livelihood, their gender, their access to information, their housing, and their geographic location, among many other variables.

Importantly, climate change has been recognized as a "poverty multiplier," whereby those who already suffer from adverse life circumstances will suffer the greatest losses from environmental changes, leading to a widening of health inequities, particularly in developing countries (17, 18).

Globally, over 3,3 billion people, mostly located in low-resource settings, are subjected to a higher magnitude and spread of anthropogenic ecosystem disruptions, such as extreme meteorological events, while simultaneously having less adaptive power when vital ecosystem services are compromised (19). Countries of the Global South are responsible for less than 8% of excess emissions (20), yet they are the most exposed and the least able to adapt to climate risks (see Figures 2 and 3). Similarly, pollution, mainly stemming from high consumer demand in high-income countries, is responsible for approximately 9 million deaths per year, of which 92% are in lower and middle-income countries (21).

Specifically, many lower power groups who are adversely impacted by environmental disruptions are worth highlighting. The WHO emphasizes women, children, ethnic minorities, communities living in poverty, migrants or displaced persons, older populations, and those with underlying health conditions as the most disadvantaged and vulnerable groups (24). For instance, extreme heat disproportionately affects pregnant people, children, and the elderly, all of whom have a higher physiological vulnerability to heat. Heat risk is then exacerbated for those people living in low-resource settings without appropriate infrastructure and technology to protect them from dangerous heat (25). Furthermore, people who live alone and those who have lower education levels and income are less likely to adopt measures to protect themselves from extreme weather events and often reside in areas where such events occur in higher frequency and magnitude (26). On the other hand, Indigenous Peoples and rural dwellers also tend to have a more direct reliance on proximity ecosystem services, putting them at higher risk of suffering adverse health outcomes from minor ecosystem disturbances (27). For instance, Indigenous peoples in the Canadian North rely on stable meteorological patterns to maintain their traditional lifestyles, including housing, hunting, and transportation habits. As such, a higher rate of warming in the North combined with a higher reliance on ecosystem stability is already affecting the livelihoods of those Indigenous People, while their ability to protect their health from environmental risks remains limited (28). Similarly, the livelihoods of over fifty percent of workers across the Asia-Pacific and Sub-Saharan Africa regions depend on agriculture. These people are already experiencing food insecurity and income losses due to climate-induced decreases in agricultural yield (17). Overall, lower power groups are not just at more risk of health losses in the near and distant future but are already suffering from the consequences of global environmental changes, delaying or even regressing their opportunities for progress towards better lives.

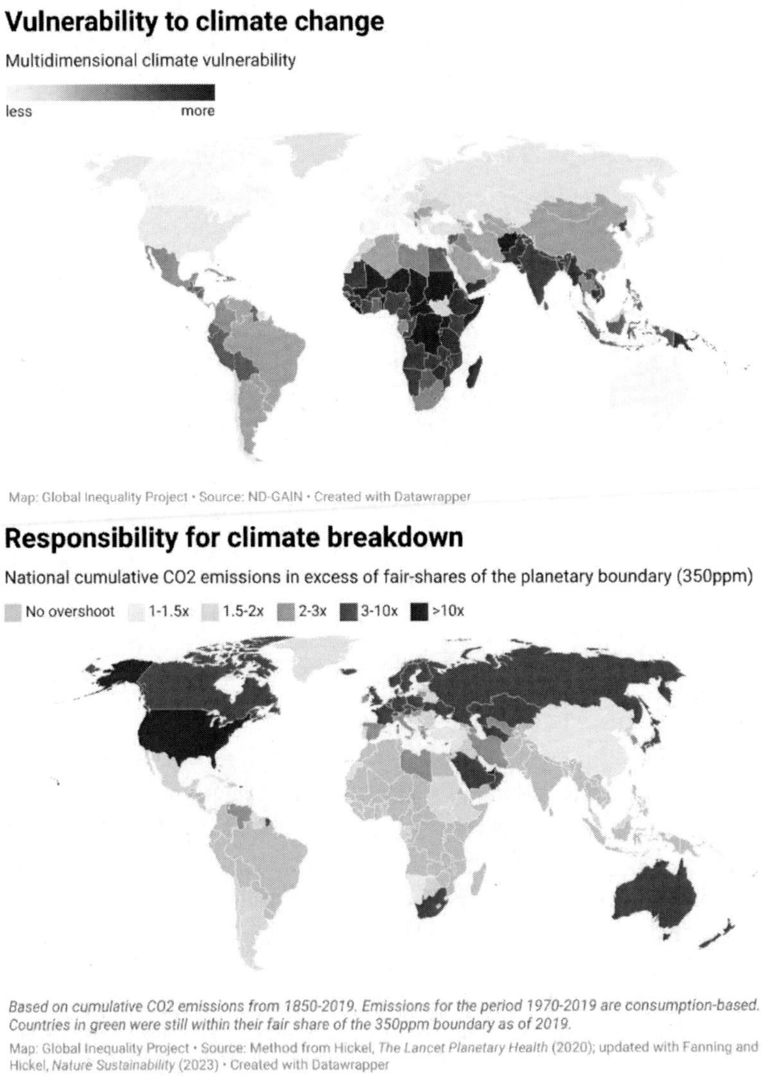

Figure 3. Overshoot greenhouse gas (GHG) emissions and climate vulnerability per country. Author: Jason Hickel. Source (22) with data from (20, 23). Reproduced with permission from Jason Hickel.

Observation 2: Those Who Are More Affected by Environmental Change Have Less Power to Act on It

In August 2022, thirty-three million people could not prevent disaster from striking their homes, as unprecedented flooding devastated Pakistan. These people - and many more around the world - are facing the risks of hunger, infectious diseases, as well as physical and mental traumas due to extreme environmental changes they did not cause (29).

Figure 4. Floods in Pakistan, 2022 © Ali Hyder Junejo, CC BY 2.0 (30).

As expressed in box 1, the defining characteristic of power is the capacity, or lack thereof, to achieve or influence a desired outcome, even in the face of opposition. Such capacity is lacking for most individuals and collectives who are currently impacted by environmental stressors. The ability of an individual to influence global and local decision-making processes is hindered by an intersection of factors, including gender, race/ethnicity, income level, access to technology, and education level, among many other variables within the power matrix. Even states can have a limited capacity to influence global policy due to external imbalances in political capital, internal economic barriers, and surreptitious influence through non-state actors (31).

Globally, most decisions that impact Earth's natural systems and, subsequently, people's health and wellbeing, take place at the multilateral, large nation-state, and multinational corporate levels, where high-power entities often disregard the interests of billions of people who may be affected by their decisions. Global treaties, fora, and multilateral bodies, such as the Paris Agreement emerging from the UNFCCC, the Convention on Global Biodiversity, and the Intergovernmental Science-Policy Platform on Biodiversity and Ecosystem Services (IPBES), are where many of the decisions regarding the global environment are taken. Trade-related organizations, such as the World Trade Organization, closed clusters of nation-state groups, such as G7, G20, BRICS, and private investment conglomerates and corporations, also influence current and future trends of resource extraction, global consumption, waste generation, and underlying supply chains. Decisions made in these exclusive spaces determine the state of Earth's natural systems along with many eco-social determinants of health and their impact on populations worldwide (32). Within decision-making bodies, the most powerful actors, such as large nation-states, multinational corporations, and hegemonic groups have the most power and influence and tend to prioritize their own interests (33–35). Although multilateralism is built around the "egalitarian" principle of one country, one vote (36), global governance remains pseudo-democratic as a small number of highly powerful individual actors can outweigh the influence of many lower power groups who are more likely to be impacted by consequent global environmental change. Granted, less developed states have helped steer global discussions towards the adoption of a more ambitious 1.5-degree temperature rise target in the Paris Agreement (37), they have however, not been able to spur global action at the required level of ambition to meet those targets, that is critically necessary to protect vulnerable citizens from disproportionate impacts of global environmental change (38).

Within global decision-making for population groups who are the most affected by environmental disruptions remain underrepresented, although their perspectives are vital to informing sustainable actions. Gender disparities are a striking example. Women, who are more vulnerable to climate risks, constitute half of the global population, yet they represent less than one-third of decision-makers in six out of nine global decision-making processes such as the Conferences of Parties (COPs), according to a survey performed in 2015 (39). Similarly, women represent less than one-third and one-fourth of National Focal Points of the Global Environmental Facility and of the UN Forum on Forests, respectively. Ministerial-level gender gaps are even more striking. Only 12% of Ministers across national environmental-sector ministries are women (39). Notably, those women who access decision-making spheres are likely to have been empowered through favorable socioeconomic determinants (40)- i.e., they are more educated, earn independent income, etc. - and thus seldom represent the most vulnerable women who suffer the most from environmental change.

At local levels, within nation-states and regions, non-consenting lower power groups are forced to endure hostile environmental conditions due to higher power interests (41). Lower power collectives are often actively discouraged from engaging in sustainability decisions that directly impact their lives. For instance, Indigenous Peoples, despite being disproportionately affected by environmental disruptions, remain scarcely included in climate policy and adaptation discussions, if at all (28). Additional examples include the forced displacement of rural populations for large-scale energy projects that benefit large urban settings (e.g., Panama, China, Peru), the rolling back environmental policies in place that protect vulnerable populations (USA), and the permissibility of large-scale deforestation for the benefit of multinational agribusiness and mining corporations (Indonesia, Brazil, Congo) (42).

Despite those challenges, many lower power groups have taken matters into their own hands and are, through movement-building, attempting to influence global and local policies. Examples include youth groups such as Fridays for Future, many national and international Indigenous groups, and groups representing Small Island Developing States (43). While they have played a crucial role in raising public, political, and corporate support for a just ecological transition (44), many grass-roots movements led by lower-power groups still fall short. Some have lacked representativity of the most disempowered, such as the Fridays for Future movement, which has mobilized millions of young urbanites, but rural youth at high risk of global environmental change have found little opportunity to participate (45). Many movements can be seen as tokenistic, decorative approaches to real representation leading to low-impact participation. Most significantly, low-power groups remain ostracized, if not demonized, in the public arena for promoting alternative worldviews that clash with hegemonic paradigms regarding our relationship with nature, individual and social responsibilities, lifestyles and consumption habits, to name a few. In many parts of the world, lower power groups suffer strong repression by law-enforcement authorities who tend to act according to the interests of higher power groups, perpetuating historical power dynamics rooted in discrimination and colonialism (46). Some of the most striking examples are seen in the violent repression of Indigenous pro-environmental pacific protests in Canada, the USA, and Latin America (47), as well as in the hundreds of environmental activist deaths annually, mainly in the most disadvantaged communities worldwide (48,49). Despite some laudable efforts to empower vulnerable groups through mechanisms such as affirmative action, gender parity laws, and reparations, their ability to instigate ambitious sustainable solutions remains marginal even when they are invited to participate in decision-making processes through formal or informal means.

Observation 3: Lower Power Groups Can Suffer Disproportionate Adverse Consequences from Actions Addressing Global Environmental Changes

> For the Zenú and Emberá Indigenous people in Colombia, the global rush for electric vehicles is bad news: nickel mining for electric batteries is contaminating their land and making their children sick (50).

Many of the proposed solutions to address the ecological crisis place an excessive burden on lower-power populations. Fundamentally, this might be due to the lack of involvement of lower power groups in the planning of interventions, resulting in actions that do not always address their needs and may lack local and cultural contextualization. Additional mechanisms explaining this are still being elucidated but could be related to lower capacity to adapt to new policies, lower access to resources such as financial means, and slow uptake of new technologies. As a result, some sustainability efforts might actually be exacerbating preexisting power and health inequities.

First, some current climate policies can disproportionately affect lower power groups. Carbon taxes are an example. In most countries, due to high-income inequality, low-income households often bear a relatively larger burden of the carbon tax than higher-income households, who can better absorb the costs of necessities such as transportation and heating affected by the carbon tax. Recent upheavals, such as the Yellow Vest movement in France have shown that there is a limit to what disadvantaged groups can tolerate. Policies that fail to include necessary measures to relieve the burden from lower power groups not only risk widening existing inequities but also undermine the social and political acceptability of effective- and urgently necessary- climate policies (51,52). Forest conservation and biodiversity protection can also have similar consequences. Case studies in Vietnam and Tanzania show that small rural communities are being given the responsibility of conservation efforts, many times shifting traditional ways of life without giving appropriate alternatives for new livelihoods (53). Overall, carbon mitigation strategies that fail to integrate equity and local needs related to energy transition and future energy consumption trends can increase and perpetuate health and wellbeing inequities between power groups (54).

Second, "green" technologies embedded in the decarbonization strategies of higher-power groups often outsource environmental and health risks to lower-power areas. Many technologies labeled as "sustainable" in fact, require an intensification of resource extraction that harms lower-power communities. Nickel is thought to be the most important "transition mineral": it is a key component of electric batteries, solar panels, and wind turbines underlying low-carbon energy and transportation. Paradoxically, nickel mining is already creating "green sacrifice zones," mainly located in impoverished communities (55). People who live in these zones are exposed to heavily polluted air, water, and soil from mining activities, leading to health problems, undermining local food sources, and threatening fragile biodiverse ecosystems. Nickel mining has also led to numerous human rights violations and conflicts, particularly affecting Indigenous Peoples across the Global North and South alike. In Papua New Guinea, the dumping of mine waste into waterbeds by a nickel battery plant has generated severe distress and uprising from local fishing communities. In the Philippines and Guatemala, notable local leaders and Indigenous rights activists opposed to nickel mining have been murdered (56). Similar harms to lower power communities have been associated with other "transition" minerals such as lithium and cobalt (57). On the other end of the life cycle, the inevitable disposal of green energy materials such as solar panels poses a serious environmental health challenge, potentially exposing low-resource populations near disposal sites to hazardous waste (58). Overall, many "green" transition measures put forth by higher power groups disregard the associated significant socio-ecological damages to lower power groups worldwide.

Furthermore, as the climate crisis advances, increasingly risky climate engineering solutions are being considered by higher power groups. The IPCC suggests that in order to decarbonize the future of energy use, the world must rely on negative emissions technologies, especially the use of bioenergy with carbon capture and storage (BECCS). According to Hickel and Slamersak (59): "scaling bioenergy monoculture would require large amounts of land—up to three times the size of India—with devastating effects on biodiversity, forests, water tables, and food systems," especially land being diverted from food crops to energy biocombustion crops (59–61). Another engineering approach relies on solar radiation management, for instance, with aerosol injection into the stratosphere to "cool" the Earth. The consequences of such a drastic intervention could be devastating, particularly for vulnerable ecosystems and the populations that depend on them. Yet, the adverse impacts on lower power groups are not always accounted for in climate engineering fora. Discussions are mostly restricted to a "small group of elites" (62) in corporate and governmental spheres of the high-power groups, mainly in the USA, the UK, and Germany (63). As climate engineering technologies progress, a dangerous opportunity is opening for higher power entities to force abrupt and potentially harmful meteorological changes on non-consenting lower power groups. Novel global governance mechanisms (64) are urgently required to ensure that power inequities underlying climate change do not force disaster again on powerless people through attempts to contain it.

Importantly, it is worth noting that not all actions addressing global environmental changes will have or are having negative impacts on communities with less power. A flurry of examples of positive changes can be found in the literature - see for example (65). However, without using a multidimensional power analysis, there will continue to be "winners and losers" (66) of global environmental actions. The latter are not a product of the constitutive, social, and environmental factors of the lower power populations per se, but a byproduct of ill-designed solutions in the name of a green transition. The movement towards a better future must be inclusive and just, or it will not achieve more than perpetuating the existing power and health inequities.

Observation 4: Issues Disproportionately Affecting Lower Power Groups are Insufficiently Addressed by the Global Community

> At the COP26 in Glasgow, yet another attempt to fund climate damage reparations for vulnerable nations has failed. This adds to a long list of shortcomings in addressing the needs of those who are most affected by environmental disruptions (67).

Figure 5. COP26, Glasgow © Francis McKee, CC BY 2.0 (68).

The global community, including decision-making bodies, media, corporations, and citizens, has always neglected issues that mainly affect people with less power. This neglect comes in many forms, including chronically low funding; few global and local policies addressing these challenges (21); low public awareness and attention by the global news media (69); a lack of accountability mechanisms and weak governance structures (38); low economic incentives to work on solutions; and little research efforts (70).

As a result, global environmental challenges that threaten the wellbeing of populations with less power tend to be lower on the global priority list (71). This is similar to historical patterns observed with Neglected Tropical Diseases - diseases burdening mostly communities living in poverty, women, and children in the tropics - and with the 10/90 gap, whereas 90% of the world's investment in health research addresses 10% of the global health problems (72). Likewise, current efforts are insufficient to protect human health from the greatest environmental anthropogenic risks, such as increasing temperatures, sea level rise, flooding, pollution, and extreme weather events, which are, for now, disproportionately impacting lower power groups (17, 73).

Sea-level rise, which threatens the existence of Small-Island Developing States and the lives of nearly 3% of the global population, remains a chronically underfunded issue of global concern (74). Threats that are considered more "acute" are prioritized over "slow-onset threats," such as sea-level rise, even when this issue was framed as a potential global security concern of high importance due to its potential to cause forced population displacement and conflict (43,75).

Similarly, unprecedented flooding in Bangladesh, Pakistan, and other South Asian states remain without adequate relief funding from high-emitting nations that are most responsible for climate change. As a result, millions of already poor families who contributed least to the climate crisis have to finance disaster recovery and climate adaptation themselves (74,76). The total price paid by low-income households in Bangladesh amounts to nearly 2 billion USD annually, which is twice the government spending and twelve times international donors' contribution to climate adaptation (77). Despite such stark inequity, the highest emitting nations, including the European Union States and the USA, continue to actively resist calls to substantially fund climate adaptation efforts in poor countries (67).

Furthermore, pollution is recognized as one of the most important risk factors for morbidity and mortality globally, with a substantial death toll in Low and middle-income countries (LMICs) (78). Yet, efforts to reduce it are scarce in official development assistance, global governance mechanisms, and global philanthropy. Air pollutant levels in LMICs still surpass WHO thresholds many fold, and less than 1% of cities in LMICs respect WHO air quality standards, compared with 17% of cities in high-income countries (HICs) (78). Inequities in exposure to air pollution are also observed at a local level. In the United States, air pollution remains inequitably distributed despite the improvement in regional air quality across the country. Communities of color and those with lower educational attainment or income levels still experience higher pollution levels and associated health risks (79). Several years of air quality improvements across the US have not solved existing disparities: the least and most polluted census tracts remained the same between 1981 and 2016 (80).

Observation 5: Higher Power Groups Are Delaying Action to Address the Health Emergency Caused by Global Environmental Change

> In 2019, a day after declaring a national climate emergency, Canada approved an oil pipeline expansion project (81).

Figure 6. Trans-mountain pipeline construction with Rockies backdrop©Adam Jones , CC BY 2.0 (82).

Many similar contradictions worldwide reflect the power imbalances underlying climate inaction: those who have the power to drive the "rapid, far-reaching and unprecedented changes in all aspects of society" required to avoid catastrophic impacts of environmental change on health (19) instead, continue to reinforce the status quo. Our main points addressed above (Observations 1 through 5) might explain this delay to act. A number of authors have pointed to possible underlying causes such as short term planning,, predominant self-interest in international politics, a lack of empathy, discriminatory attitudes, weak and unaccountable governance structures, and the drive for economic growth above all else, among others could be to blame (83–85). Twenty years of climate negotiations and mounting evidence of the health impacts of environmental change have failed to accelerate environmental action in higher power groups sufficiently to curb the health and environmental emergency. The power imbalance underlying slow environmental action manifests in several interlinked ways.

First, higher power entities are not acting quickly enough to keep global warming below 1.5 degrees and limit environmental degradation. According to the UNFCCC and the WHO less than ten years are remaining to prevent environmental disruptions that have the potential to reverse fifty years of hard-earned progress in human wellbeing and exacerbate health inequities worldwide (24). The pace of change towards the so-called Great Transition - a restructuring of our social and economic systems (86) - is still too slow (see Figure 4). There are still no climate action plans that meet the goals of the Paris Agreement (87). As of 2022, mainly symbolic "climate emergency" declarations have been adopted by eighteen countries and the European Union (88). Many countries have been waiting to follow the ambition level set by the most developed and highest emitting countries. Slow action by the most powerful states has thereby stalled efforts relying on global coordination: with the exception of the United Kingdom and Norway, the most powerful and highest emitting countries still have less ambitious action plans than smaller powers such as Costa Rica, Ethiopia, Nepal and Morocco (89). Sector-specific commitments, such as net-zero health systems at the COP26 Health program, have mainly rallied lower-income countries (90). More powerful states are far from their "fair share" contribution, as the top three emitters, China, the USA, and the EU, still contribute sixteen times the emissions of the lowest 100 countries (91) and keep resisting efforts to pay financial reparations for countries who are already suffering from environmental changes. The USA delayed ambitious climate action for a decade, although the latest adoption of the "Inflation Reduction Act" has shed light on renewed ambition for national and global efforts (92). Beyond climate change, other environmental changes impacting human health and wellbeing, such as biodiversity loss and air pollution, remain stalled by industrial and corporate interests. In fact, major oil and gas companies are heavily investing in climate disinformation, using "greenwashing" initiatives to distract the public from recognizing their devastating effects on global climate and ecosystems (93). As a result, there are early signs that we have already reached tipping points for some of our environmental indicators (54, 94), while irreversible damage is already present in many ecosystems and to biodiversity (95).

Second, the power balance still favors those who are resisting ambitious environmental action. A few highly powerful individuals have stalled global efforts toward a just transition. For instance, it was very easy for the Trump administration to retract the USA from the Paris Agreement despite knowing that climate change will disproportionately affect people of color across the US. This decision has since been reversed by the Biden administration, which, however, continues to advance fossil-fuel projects across the country under the pressure of oil and gas lobbies (96, 97). Similarly, Jair Bolsonaro influenced Brazil's reluctant stance to advance the Convention on Biological Diversity (CBD) agenda despite its massive support by Brazilian Indigenous Groups who rely heavily on the Amazon rainforest (98). Moreover, increasingly powerful actors, including billionaires, celebrities, and political figures, are portraying themselves as environmental "activists" while perpetuating the status quo by fueling the ecological crisis with their resource-intensive lifestyles and corporate activities, failing to model the transformational changes needed to respect planetary boundaries (99). On other fronts, good-intentioned individuals and groups in relatively privileged positions of society are trying to tip the power balance towards environmental action (100). Healthcare professionals and organizations are an example: they are increasingly using their power as trusted messengers of science to push for ambitious climate action within and beyond the healthcare sector. The World Health Organization has declared that climate change is "the greatest threat to global health in the 21st century" (101). Around the world, healthcare activism has won several battles, such as achieving coal phase-out in Canada, reducing healthcare's ecological footprint in the UK, and spurring global commitments to low-carbon, resilient health systems (102, 103). Ahead of COP26, unprecedented mobilization around the Healthy Climate Prescription rallied 600 organizations representing millions of healthcare workers worldwide (100). Despite their relatively higher power as compared with other civil-society activist groups, healthcare professionals have yet to overcome the stronger influence of industry lobbies that stalls global and national environmental agendas (104).

Fundamentally, reactions to global environmental change prioritize issues that affect populations with more power and/or relate to their own interests (69). Climate action remains dependent on political will (19), which, even in democratic societies, mainly prioritizes short-term governmental goals and corporate interests over citizens' best interests in the long term. The UK's recent decision to lift a ban on fracking for gas shows that even the most ambitious climate action plans are at the mercy of changing political will, influenced by industrial lobby groups (105). Higher power entities are not only often inactive, many also actively promote the status quo - whether intentionally or unintentionally - by perpetuating a consumption-based global culture that reinforces the predominant neoliberal (i.e., deregulation of the economy, minimal state intervention, reduction of social services by the state) and neocolonial institutional order (83). Not only is this current order driving health inequities worldwide (106), but its objective of perpetual growth is also incompatible with the finite boundaries of the planet (107). To this day, no powerful entities have pushed for alternative economic systems in an equal measure to balance planetary health with economic interests. What seems to be evident is that despite global calls for equity and inclusion, populations with more power must still do much more to avoid mortgaging *"the health of future generations to realize economic and development gains in the present"* (108).

Observation 6: As the Degree of Environmental Change Increases, Everyone, Including Higher Power Groups, Suffers Adverse Health Impacts

> Celebrities' homes burning in California's 2018 wildfires made it clear that no one is safe from environmental disasters. We are currently at an inflection point in the stability of Earth's natural systems. Maintaining the status quo risks destabilizing Earth's natural systems and creating a less hospitable Earth that would reverse the gains in human wellbeing accomplished over the last two hundred years of global progress (5).

At this point, the wellbeing of all population groups, regardless of their position on the power matrix, is likely to be further threatened as planetary boundaries are surpassed, and tipping points are reached. There is still no definitive consensus on where the limits of the Earth's stability are, as we are still trying to grasp the full complexity and the intrinsic uncertainties of Earth Systems (94, 109). So far, the Planetary Boundaries framework (5) is the largest global effort to delineate a safe operating space for humanity. Surpassing the planetary boundaries could be potentially devastating: it could mean a reversal of the environmentalist paradox for all groups, regardless of their power status. The rate at which this change might come about is still uncertain and is related to systems inertia as well as to system tipping points (109). A tipping point in an Earth system is a critical point at which changes in said system happen in a non-linear fashion and convey a high risk of adverse consequences (110). Tipping points are projected to be irreversible and may cascade into other tipping points of interconnected Earth systems (94, 111). For example, the weakening of natural carbon sinks, such as large ice sheets in the polar regions, might quickly destabilize other Earth systems like ocean currents (111). In this sense, the portion of the graph (Figure 1) to the right of "tipping points are reached" is plotted with a high grade of uncertainty. Social tipping points are also to be considered in rapidly declining human wellbeing. Environmental impacts have the potential to push social systems into a state of instability, exacerbating issues like political and economic fragility that, through amplifying loops augment the risk of displacement, conflict, and food insecurity (112).

Figure 7. California wildfires: The holy fire at Lake Elsinore on August 9, 2018. ©Kevin Key, CC BY-NC-SA 2.0 (113).

Recent events have shown that the consequences of global environmental change are becoming more difficult to avoid, even for higher-power population groups. Environmental disruptions of higher frequency and magnitude are increasingly reaching high-income nations, exposing more people who are in positions of power to environmental changes. Middle and high-income families *have* been hit by cyclones and hurricanes in coastal areas worldwide, wildfires in California, flooding in the UK, and heatwaves across Europe, to name a few recent events (114–118). Unmitigated climate change is predicted to cause stronger and more frequent hurricanes and cyclones (among other extreme weather events), which are expected to inflict unprecedented losses to high-income communities and economies worldwide in the next decades (119). As disruptions spread widely and indiscriminately across inhabited areas in North America and Europe, they are hitting closer to the homes of more privileged populations. This may help them to become aware and empathize with other less privileged people who suffer from similar events elsewhere in the world or in other communities at home (120). Moreover, some anthropogenic environmental changes, such as air pollution, cause chronic health risks for all population groups, even at low exposure levels. Nearly 100% of city dwellers across Europe are exposed to harmful pollutant levels that surpass WHO limits, causing morbidity and mortality even among economically and socially privileged groups (121). As a result, communities with more power also inevitably suffer health losses from environmental change, even though it may be to a lesser degree than communities with less power.

Overall, although there is still little high-certainty evidence indicating how environmental changes will progress or at what rate these will happen (94), similar to authors like David Wallace-Wells (122), we find that it is highly unlikely that any group of humans will be able to sustainably thrive in an inhospitable world in which cascading natural disasters, large populations displacement and widespread food insecurity are ever more common. While some population groups may adapt better to social and environmental disruptions, it will become increasingly difficult for higher power groups to shield themselves from the social and environmental instability caused by global environmental change. It is unclear if this would mean a total reversal of inequities (as suggested in Figure 1), or if some groups would be able to adapt to drastically different biosphere conditions than that of the Holocene. The point to be made is that our shared common existence/wellbeing as humans is bound to our shared planetary home and the living systems we depend on. All groups, all nations, and all people must recognize this reality and accelerate action toward the planetary health agenda.

Moving Forward

Currently, human civilization sits at a decision impasse. Never before, with the exception of current and past threats of nuclear wars that would assure global destruction, have we been at a bottleneck in which the decisions we make today will have long-lasting ripple effects for generations to come. The last fifty years of global development efforts have undoubtedly improved the health of people around the world, yet they have perpetuated health and economic inequities through discriminatory, post-colonial power dynamics. Resource over-extraction has eroded the stability of ecosystems on which billions of people depend. Population groups with more power are benefiting more from the current resource-intensive global economic model at the expense of the health and wellbeing of people with less power. Those who are least responsible for global environmental change are those who suffer the most from it while also having less power to act on it. In the current status quo, environmental change—and efforts to address it—are worsening underlying health and power inequities.

This current state of affairs undermines the ethical foundations of our shared humanity: insufficient action to address global environmental change condemns billions of us to extreme suffering caused by a preventable anthropogenic phenomenon. We must recognize the longstanding power and health inequities that are impeding us from building resilient and just societies and from achieving Sustainable Development Goals (16–18). To be effective, all efforts to address global environmental change must consider the question: what causes health and power inequities in the first place? Viewing health and environmental inequities through a power lens helps us identify key structural mechanisms which must be challenged in order to curb environmental changes and improve health at once. This includes questioning the global economic and political system that underlies health inequities worldwide.

We argue that a paradigm shift is necessary in the global socio-political order to favor economic and global governance models that respect the ecological limits of the planet and protect human wellbeing regardless of power status. Leading economists are proposing alternative economic paradigms that would allow reducing inequities and respecting ecosystems needed for our health (85, 123). A multiplicity of evidence-based just transition solutions adapted to local and global contexts have been proposed (124). In fact, many solutions to the ecological crisis lie at the intersection of power, health, and environmental change (83). We note, however, that more empirical evidence is needed to understand what the global health implications of paradigms such as degrowth and others would be.

Moving forward, there is an imperative to understand global environmental change as a complex social challenge, not only a technical one. An intersectional power lens is urgently needed in research, policymaking, and across all other efforts that aim to address global environmental change.

Literature still remains limited in its definition and evidence-based measurement of power, and there is a need to quantify and characterize power in its multiple dimensions and with an intersectional approach. In that process, it is necessary to advance transdisciplinary research and solutions that integrate not just different disciplinary fields but incorporate different knowledge holders, including those who are most affected by global environmental change. Those in power and in privileged positions in society, including not only policymakers but also researchers, funders, entrepreneurs, and practicing professionals, need to "lean out" and allow the increased collaboration with lower power groups to inform and support the generation of social movements that can effectively shift power dynamics towards prioritizing environmental justice. Part of the answers might entail looking to our past and understanding the levers of social change that informed transformations achieved through social movements that were successful in achieving long-lasting policy changes, such as abolishing slavery, voting rights for women, and HIV treatment access, among others. There is also significant untapped potential in mobilizing groups of a relatively higher power, such as health professionals, who can help shift the power balance towards ambitious environmental action. Another portion of the solutions will entail completely redesigning our social and economic order (85) through a process known as the Great Transition (86).

Finally, we posit the importance of reinforcing and expanding democratic and human-rights-based decision-making processes and institutions. Challenging the environmental status quo requires empowering those who suffer from global environmental change to get involved in global and local decision-making toward sustainability. Those who are suffering consequences should be given a place in the influence spheres. We must remember that "inequality is a political choice" (18) and that health inequities and ecological crises are rooted in a fundamental question: what values should prevail in times of crisis? It might be time to realign our global ethical compass.

Acknowledgments

We declare no competing interests. We acknowledge the diverse group of people and communities of practice that have influenced our analysis, especially those that continue to work for change for the greater good and that have continued to be marginalized from formal institutional settings.

References

[1] Pinker S. Enlightenment now: The case for reason, science, humanism, and progress. London: Penguin Books, 2019.

[2] Davenport R. Patterns of death, 1800–2020: Global rates and causes. In: The Routledge history of death since 1800. London: Routledge, 2021:21-44.

[3] Roser M, Ritchie H. Hunger and undernourishment. Our world in data, 2019. URL: https://ourworldindata.org/hunger-and-undernourishment.

[4] Roser M, Ortiz-Ospina E. Global education. Our world in data, 2016. URL: https://ourworldindata.org/global-education.

[5] Steffen W, Richardson K, Rockström J, Cornell S, Fetzer I, Bennett E, et al. Planetary boundaries: Guiding human development on a changing planet. Science 2015;347(6223):1259855. doi: 10.1126/science.1259855.

[6] Raudsepp-Hearne C, Peterson GD, Tengö M, Bennett EM, Holland T, Benessaiah K, et al. Untangling the environmentalist's paradox: Why is human well-being increasing as ecosystem services degrade? BioScience 2010;60(8):576–89. doi: 10.1525/bio.2010.60.8.4.

[7] Weber M. The theory of social and economic organization. New York: Simon Schuster, 2009.

[8] Bell K. Agency. Open education sociology dictionary, 2016. URL: https://sociologydictionary.org/agency/.

[9] UNICEF. The climate crisis is a child rights crisis. UNICEF, 2021. URL: https://www.unicef.org/reports/climate-crisis-child-rights-crisis.

[10] UNICEF. Children are facing deadly drought in the horn of Africa. UNICEF, 2022. URL: https://www.unicef.org/stories/climate-drought-horn-of-africa.

[11] Ayene M. Somali region drought response. Flickr, 2022. URL: https://www.flickr.com/photos/unicefethiopia/51859324308/.

[12] Shiva V. Oneness vs. the 1%: Shattering illusions, seeding freedom. White River Junction, VT: Chelsea Green Publishing, 2020.

[13] Smit B, Wandel J. Adaptation, adaptive capacity and vulnerability. Glob Environ Change 2006;16(3):282–92. doi: 10.1016/j.gloenvcha.2006.03.008.

[14] Simpson NP, Mach KJ, Constable A, Hess J, Hogarth R, Howden M, et al. A framework for complex climate change risk assessment. One Earth 2021;4(4):489–501. doi: 10.1016/j.oneear.2021.03.005.

[15] Giordano F. Climate change vulnerability and risk - key concepts. Istituto superiore per la protezione e la ricerca ambientale, 2014. URL: http://www.lifesecadapt.eu/fileadmin/user_upload/ALLEGATI_LIFESECADAPT/documenti/Vulnerability_Risk_FGiordano.pdf.

[16] Georgieva K, Gaspar V, Pazarbasioglu C. Poor and vulnerable countries need support to adapt to climate change. IMF, 2022. URL: https://www.imf.org/en/Blogs/Articles/2022/03/23/blog032322-poor-and-vulnerable-countris-need-support-to-adapt-to-climate-change.

[17] Sarkodie SA, Ahmed MY, Owusu PA. Global adaptation readiness and income mitigate sectoral climate change vulnerabilities. Humanit Soc Sci Commun 2022;9:113. doi: 10.1057/s41599-022-01130-7.

[18] Chancel L, Piketty T, Saez E, Zucman G. World inequality report 2022. World inequality lab, 2021. URL: https://wir2022.wid.world/www-site/uploads/2021/12/WorldInequalityReport2022_Full_Report.pdf.

[19] IPCC. Summary for Policymakers. In: Masson-Delmotte V, Zhai P, Pörtner HO, Roberts D, Skea J, Shukla PR, eds. Global warming of 1.5°C. An IPCC special report on the impacts of global warming of 1.5°C above pre-industrial levels and related global greenhouse gas emission pathways, in the context of strengthening the global response to the threat of climate change, sustainable development, and efforts to eradicate poverty. Geneva: World Meteorological Organization, 2018.

[20] Hickel J. Quantifying national responsibility for climate breakdown: An equality-based attribution approach for carbon dioxide emissions in excess of the planetary boundary. Lancet Planet Health 2020;4(9):e399–404. doi: 0.1016/S2542-5196(20)30196-0.

[21] Fuller R, Landrigan PJ, Balakrishnan K, Bathan G, Bose-O'Reilly S, Brauer M, et al. Pollution and health: a progress update. Lancet Planet Health 2022;6(6):e535–47. doi: 10.1016/S2542-5196(22)00090-0.

[22] Hickel J. The tragedy in Pakistan is a stark reminder of climate injustice. Twitter, 2022. URL: https://twitter.com/jasonhickel/status/1566387987235176449.

[23] University of Notre Dame. ND-GAIN. Notre Dame Global Adaptation Initiative, 2022. URL: https://gain.nd.edu/.

[24] World Health Organization. Climate change and health. Geneva: WHO, 2021. URL: https://www.who.int/news-room/fact-sheets/detail/climate-change-and-health.

[25] Conlon KC, Mallen E, Gronlund CJ, Berrocal VJ, Larsen L, O 'Neill Marie S. Mapping human vulnerability to extreme heat: A critical assessment of heat vulnerability indices created using principal components analysis. Environ Health Perspect 2022;128(9):097001. doi: 10.1289/EHP4030.

[26] Thomas K, Hardy RD, Lazrus H, Mendez M, Orlove B, Rivera-Collazo I, et al. Explaining differential vulnerability to climate change: A social science review. Wiley Interdiscip Rev Clim Change 2019;10(2):e565. doi:10.1002/wcc.565.

[27] Celidwen Y, Schultz C, Horn O, Githaiga C, Vera M, et al. The determinants of planetary health: An Indigenous consensus perspective. Lancet Planet Health 2021;5(3):e156–63. doi:10.1016/S2542-5196(21)00354-5.

[28] Reed G, Gobby J, Sinclair R, Ivey R, Matthews HD. Indigenizing climate policy in Canada: A critical examination of the pan-canadian framework and the ZéN roadmap. Front Sustain Cities 2021;3:644675. doi: 10.3389/frsc.2021.644675.

[29] Human Rights Watch. Pakistan floods dire for millions of children. Human Rights Watch, 2022. URL: https://www.hrw.org/news/2022/09/07/pakistan-floods-dire-millions-children.

[30] Junejo AH. Flood in Pakistan 2022. Flickr, 2022. URL: https://www.flickr.com/photos/193804179@N08/52331043544/.

[31] Hoffman SJ. Mitigating inequalities of influence among states in global decision making. Glob Policy 2012;3(4):421–32. doi: 10.1111/j.1758-5899.2011.00153.x.

[32] Willetts L, Bansard J, Kohler P, Bettelli P, Rosen T, Schröder M, et al. Health in the global environmental agenda: A policy guide. International Institute for

Sustainable Development, 2022. URL: https://policycommons.net/artifacts/2211433/health-in-the-global-environmental-agenda/2968229/.

[33] Clapp J. The problem with growing corporate concentration and power in the global food system. Nat Food 2021;2(6):404–8.

[34] Meagher M. Competition is killing us: How big business is harming our society and planet - and what to do about it. London: Penguin Books, 2020.

[35] Stuckler D, King L, Robinson H, McKee M. WHO's budgetary allocations and burden of disease: a comparative analysis. Lancet 2008;372(9649):1563–9. doi: 10.1016/S0140-6736(08)61656-6.

[36] United Nations. UN Charter. New York: United Nations, 2022. URL: https://www.un.org/en/about-us/un-charter.

[37] Brun A. Conference diplomacy: The making of the Paris agreement. Polit Gov 2016;4(3):115–23. doi: 10.17645/pag.v4i3.649.

[38] Climate Action Tracker. Home. Climate Action Tracker, 2022. URL: https://climateactiontracker.org/.

[39] International Union for Conservation of Nature. EGI: Women's participation in global environmental decision-making factsheet. IUCN, 2015. URL: https://genderandenvironment.org/egi-womens-participation-in-global-environmental-decision-making-factsheet/.

[40] Ventura AM, Morillas L, Martins-Loução MA, Cruz C. Women's empowerment, research, and management: their contribution to social sustainability. Sustainability 2021;13(22):12754. doi: 10.3390/su132212754.

[41] Schroeder H. Agency in international climate negotiations: the case of Indigenous peoples and avoided deforestation. Int Environ Agreem Polit Law Econ 2010;10(4):317–32. doi: 10.1007/s10784-010-9138-2.

[42] EJatlas, Mapping environmental justice. The global atlas of environmental justice, 2022. URL: https://ejatlas.org/.

[43] van Schaik L, Sarris S, von Lossow T. Fighting an existential threat: Small island states bringing climate change to the UN Security Council. Clingendael Institute, 2018. URL: https://www.jstor.org/stable/resrep17348.

[44] Thiri MA, Villamayor-Tomás S, Scheidel A, Demaria F. How social movements contribute to staying within the global carbon budget: Evidence from a qualitative meta-analysis of case studies. Ecol Econ 2022;195:107356. doi: 10.1016/j.ecolecon.2022.107356.

[45] Barford A. Climate talks will fail without more young people's voices. Here's why. World Economic Forum, 2021. URL: https://www.weforum.org/agenda/2021/10/why-climate-change-summits-need-young-peoples-voices/.

[46] UN Press. Indigenous peoples disproportionately impacted by climate change, systematically targeted for defending freedoms, speakers tell permanent forum. New York: United Nations, 2018. URL: https://press.un.org/en/2018/hr5389.doc.htm.

[47] Scheidel A, Del Bene D, Liu J, Navas G, Mingorría S, Demaria F, et al. Environmental conflicts and defenders: A global overview. Glob Environ Change. 2020;63:102104. doi: 10.1016/j.gloenvcha.2020.102104.

[48] Hines A. Decade of defiance. Global witness, 2022. URL: https://www.globalwitness.org/en/campaigns/environmental-activists/decade-defiance/.

[49] Pross C. Rising violence against environmental activists threatens us all. Stockholm Environment Institute, 2021. URL: https://www.sei.org/perspectives/violence-environmental-activists-threatens/.

[50] ABColombia. Indigenous Peoples struggle for a dignified life against giant Nickel Miner South. ABColombia, 2018. URL: https://www.abcolombia.org.uk/colombia-indigenous-peoples-struggle-for-a-dignified-life-against-giant-nickel-miner-south32/.

[51] Andersson J, Atkinson G. The distributional effects of a carbon tax: The role of income inequality. Grantham Research Institute on climate change and the environment, 2022. URL: https://www.lse.ac.uk/granthaminstitute/publication/the-distributional-effects-of-a-carbon-tax-the-role-of-income-inequality/.

[52] Kaufman N, Krause E. Putting a price on carbon: Ensuring equity. World Resources Institute, 2016. URL: https://www.wri.org/research/putting-price-carbon-ensuring-equity.

[53] Barnes J, Dove MR, eds. Climate cultures: Anthropological perspectives on climate change. New Haven, CT: Yale University Press, 2015.

[54] Le Quéré C, Andrew RM, Friedlingstein P, Sitch S, Hauck J, Pongratz J, et al. Global carbon budget 2018. Earth Syst Sci Data 2018;10(4):2141–94. doi: 10.5194/essd-10-2141-2018.

[55] Zografos C, Robbins P. Green sacrifice zones, or why a green new deal cannot ignore the cost shifts of just transitions. One Earth 2020;3(5):543–6. doi: 10.1016/j.oneear.2020.10.012.

[56] Voskoboynik DM, Farrugia JD. The nickel nexus: Mapping the frontiers of carbon neutrality. Barcelona: JHU-UPF Public Policy Center, Johns Hopkins University and Universitat Pompeu Fabra, 2022. URL: http://repositori.upf.edu/handle/10230/54138.

[57] Church C, Crawford A. Green Conflict Minerals: The fuels of conflict in the transition to a low-carbon economy. Winnipeg, MA: International Institute for Sustainable Development, 2018. URL: https://www.iisd.org/system/files/publications/green-conflict-minerals.pdf.

[58] Peplow M. Solar panels face recycling challenge. Chemical & Engineering News, 2022. URL: https://cen.acs.org/environment/recycling/Solar-panels-face-recycling-challenge-photovoltaic-waste/100/i18.

[59] Hickel J, Slamersak A. Existing climate mitigation scenarios perpetuate colonial inequalities. Lancet Planet Health 2022;6(7):e628–31. doi: 10.1016/S2542-5196(22)00092-4.

[60] Pörtner HO, Roberts DC, Tignor MMB, Poloczanska ES, Mintenbeck K, Alegría A, et al., eds. Climate change 2022: Impacts, adaptation and vulnerability. Contribution of Working Group II to the sixth assessment report of the Intergovernmental Panel on Climate Change. Cambridge, UK: Cambridge University Press, 2022.

[61] Creutzig F, Erb KH, Haberl H, Hof C, Hunsberger C, Roe S. Considering sustainability thresholds for BECCS in IPCC and biodiversity assessments. GCB Bioenergy 2021;13(4):510–5. doi: 10.1111/gcbb.12798.

[62] Winickoff DE, Flegal JA, Asrat A. Engaging the Global South on climate engineering research. Nat Clim Change 2015;5(7):627–34. doi: 10.1038/nclimate2632.

[63] Biermann F, Möller I. Rich man's solution? Climate engineering discourses and the marginalization of the Global South. Int Environ Agreem Polit Law Econ 2019;19(2):151–67. doi: 10.1007/s10784-019-09431-0.

[64] Jinnah S, Nicholson S, Morrow DR, Dove Z, Wapner P, Valdivia W, et al. Governing climate engineering: A proposal for immediate governance of solar radiation management. Sustainability 2019;11(14):3954. doi: 10.3390/su11143954.

[65] Jones IJ, MacDonald AJ, Hopkins SR, Lund AJ, Liu ZYC, Fawzi NI, et al. Improving rural health care reduces illegal logging and conserves carbon in a tropical forest. Proc Natl Acad Sci 2020;117(45):28515–24. doi: 10.1073/pnas.2009240117.

[66] Glantz MH. Assessing the impacts of climate: The issue of winners and losers in a global climate change context. In: Zwerver S, van Rompaey RSAR, Kok MTJ, Berk MM, eds. Climate change research: Evaluation and policy implications. Amsterdam: Elsevier, 1995:41–54.

[67] Rowling M. Climate "loss and damage" earns recognition but little action in COP26 deal. Reuters, 2021. URL: https://www.reuters.com/business/cop/climate-loss-damage-earns-recognition-little-action-cop26-deal-2021-11-13/.

[68] Mckee F. Cop26, Glasgow. Flickr, 2021. URL: https://www.flickr.com/photos/37825904@N06/51655095611/.

[69] Singh S. Western media is fails to cover climate change in most vulnerable countries. The Click, 2021. URL: https://theclick.news/western-media-is-failing-to-cover-climate-change-in-the-most-vulnerable-countries/.

[70] Klingelhöfer D, Müller R, Braun M, Brüggmann D, Groneberg DA. Climate change: Does international research fulfill global demands and necessities? Environ Sci Eur 2020;32(1):137. doi: 10.1186/s12302-020-00419-1.

[71] Ida T. Climate refugees – the world's forgotten victims. World Economic Forum, 2021. URL: https://www.weforum.org/agenda/2021/06/climate-refugees-the-world-s-forgotten-victims/.

[72] Tidman R, Abela-Ridder B, de Castañeda RR. The impact of climate change on neglected tropical diseases: a systematic review. Trans R Soc Trop Med Hyg 2021;115(2):147–68. doi: 10.1093/trstmh/traa192.

[73] Campbell LP, Luther C, Moo-Llanes D, Ramsey JM, Danis-Lozano R, Peterson AT. Climate change influences on global distributions of dengue and chikungunya virus vectors. Philos Trans R Soc B Biol Sci. 2015 Apr 5;370(1665):20140135. doi: 10.1098/rstb.2014.0135.

[74] Climate Finance Access Network (CFAN). Accessing climate finance: Challenges and opportunities for Small Island Developing States. United Nations, 2022. URL:

https://www.un.org/ohrlls/sites/www.un.org.ohrlls/files/accessing_climate_finance_challenges_sids_report.pdf.
[75] Martyr-Koller R, Thomas A, Schleussner CF, Nauels A, Lissner T. Loss and damage implications of sea-level rise on Small Island Developing States. Curr Opin Environ Sustain 2021;50:245–59. doi: 10.1016/j.cosust.2021.05.001.
[76] Hossain M. "Silent financier": How Bangladesh's poor are paying the costs of climate damage. Reuters, 2021. URL: https://www.reuters.com/markets/commodities/silent-financier-how-bangladeshs-poor-are-paying-costs-climate-damage-2021-12-10/.
[77] Shaikh Eskander PS. Bearing the climate burden: how households in Bangladesh are spending too much. International Institute for Environment and Development, 2022. URL: https://www.iied.org/16643iied.
[78] World Health Organization. Billions of people still breathe unhealthy air: New WHO data. World Health Organization, 2022. URL: https://www.who.int/news/item/04-04-2022-billions-of-people-still-breathe-unhealthy-air-new-who-data.
[79] Keswani A, Akselrod H, Anenberg SC. Health and clinical impacts of air pollution and linkages with climate change. New Engl J Med Evid 2022;1(7): EVIDra2200068. doi: 10.1056/EVIDra2200068.
[80] Colmer J, Hardman I, Shimshack J, Voorheis J. Disparities in PM2.5 air pollution in the United States. Science 2020;369(6503):575–8. doi: 10.1126/science.aaz9353.
[81] Mabee W. Cognitive dissonance: Canada declares a national climate emergency and approves a pipeline. The Conversation, 2022. URL: http://theconversation.com/cognitive-dissonance-canada-declares-a-national-climate-emergency-and-approves-a-pipeline-119151.
[82] Jones A. Trans-Mountain pipeline construction with Rockies backdrop - Clearwater to Valemount - BC - Canada. Flickr, 2022. URL: https://www.flickr.com/photos/adam_jones/52090931172/.
[83] Shiva V. Earth democracy: Justice, sustainability and peace. London: Zed Books, 2005.
[84] Piketty T. Capital and ideology. Cambridge, MA: Harvard University Press, 2020.
[85] Hickel J. The divide: A brief guide to global inequality and its solutions. London: Penguin Random House, 2017.
[86] Raskin P. The Great Transition today. Great Transition Initiative, 2006. URL: https://greattransition.org/file/a-great-transition-today-a-report-from-the-future-2.
[87] Climate Action Tracker. Despite Glasgow Climate Pact 2030 climate target updates have stalled. Climate Action Tracker, 2022. URL: https://climateactiontracker.org/documents/1051/CAT_2022-06-03_Briefing_MidYearUpdate_DespiteGlasgowTargetUpdatesStalled.pdf.
[88] Climate Emergency Declaration. Climate emergency declarations in 2,291 jurisdictions and local governments cover 1 billion citizens. Climate Emergency Declaration, 2022. URL: https://climateemergencydeclaration.org/climate-emergency-declarations-cover-15-million-citizens/.
[89] Climate Action Tracker. Countries. Climate Action Tracker, 2022. URL: https://climateactiontracker.org/countries/.

[90] World Health Organization. COP26 health programme. Geneva: World Health Organization, 2022. URL: https://www.who.int/initiatives/cop26-health-programme.

[91] Friedrich J, Ge M, Pickens A. This interactive chart shows changes in the world's top 10 emitters. World Resources Institute, 2020. URL: https://www.wri.org/insights/interactive-chart-shows-changes-worlds-top-10-emitters.

[92] Climate Action Tracker. USA. Climate Action Tracker, 2022. URL: https://climateactiontracker.org/countries/usa/.

[93] InfluenceMap. Big oil's real agenda on climate change 2022. Influence Map, 2022. URL: https://influencemap.org/report/Big-Oil-s-Agenda-on-Climate-Change-2022-19585.

[94] Armstrong McKay DI, Staal A, Abrams JF, Winkelmann R, Sakschewski B, Loriani S, et al. Exceeding 1.5°C global warming could trigger multiple climate tipping points. Science 2022;377(6611):eabn7950. doi: 10.1126/science.abn7950.

[95] Kumaraswamy S, Kunte K. Integrating biodiversity and conservation with modern agricultural landscapes. Biodivers Conserv 2013;22(12):2735–50. doi: 10.1007/s10531-013-0562-9.

[96] The White House. Press briefing by press secretary Karine Jean-Pierre and NSC coordinator for strategic communications John Kirby. The White House, 2022. URL: https://www.whitehouse.gov/briefing-room/press-briefings/2022/06/15/press-briefing-by-press-secretary-karine-jean-pierre-and-nsc-coordinator-for-strategic-communications-john-kirby/.

[97] Newburger E. Environmental groups sue Biden to block 3,500 oil and gas drilling permits. CNBC, 2022. URL: https://www.cnbc.com/2022/06/16/environmental-groups-sue-biden-to-block-oil-and-gas-drilling-permits.html.

[98] Gilbert N. Scientists warn deal to save biodiversity is in jeopardy. Nature 2022 Jun 30. doi: 10.1038/d41586-022-01805-w.

[99] Neubauer I. A bit rich? Billionaires' climate efforts draw scepticism, praise. Aljazeera, 2022. URL: https://www.aljazeera.com/economy/2022/5/6/a-bit-rich-billionaires-climate-efforts-draw-skepticism-praise.

[100] Healthy Climate Letter. #HealthyClimate. #HealthyClimatePrescription, 2022. URL: https://healthyclimateletter.net/.

[101] World Health Organization. WHO calls for urgent action to protect health from climate change – Sign the call. World Health Organization, 2015. URL: https://www.who.int/news/item/06-10-2015-who-calls-for-urgent-action-to-protect-health-from-climate-change-sign-the-call.

[102] Howard C. Targeted change making for a healthy recovery. Lancet Planet Health 2020;4(9):e372–4. doi: 10.1016/S2542-5196(20)30200-X.

[103] World Health Organization. Health-climate alliance launched to help countries turn commitments into action. World Health Organization, 2022. URL: https://www.who.int/news/item/27-06-2022-health-climate-alliance-launched-to-help-countries-turn-commitments-into-action.

[104] Niranjan A. Lobbying threat to global climate action. DW, 2021. URL: https://www.dw.com/en/lobbying-threat-to-global-climate-action/a-59726541.

[105] Thomas N, Pickard J. Liz Truss pledges fracking drive in England to boost energy security. Financial Times, 2022. URL: https://www.ft.com/content/58eb6f5e-a035-4749-9508-f3e9602d46f1.

[106] Kentikelenis A, Rochford C. Power asymmetries in global governance for health: a conceptual framework for analyzing the political-economic determinants of health inequities. Glob Health 2019;15(1):70. doi: 10.1186/s12992-019-0516-4.

[107] Jackson T. Prosperity without growth: Foundations for the economy of tomorrow. 2nd ed. London: Routledge, 2016.

[108] Whitmee S, Haines A, Beyrer C, Boltz F, Capon AG, Dias BF de S, et al. Safeguarding human health in the Anthropocene epoch: report of The Rockefeller Foundation–Lancet Commission on planetary health. Lancet 2015;386(10007):1973–2028. doi: 10.1016/S0140-6736(15)60901-1.

[109] Haines A, Frumkin H. Planetary health: Safeguarding human health and the environment in the Anthropocene. Cambridge, UK: Cambridge University Press, 2021.

[110] Lenton TM, Held H, Kriegler E, Hall JW, Lucht W, Rahmstorf S, et al. Tipping elements in the Earth's climate system. Proc Natl Acad Sci 2008;105(6):1786–93. doi: 10.1073/pnas.0705414105.

[111] Cai Y, Lenton TM, Lontzek TS. Risk of multiple interacting tipping points should encourage rapid CO_2 emission reduction. Nat Clim Change 2016;6(5):520–5. doi: 10.1038/nclimate2964.

[112] Franzke CLE, Ciullo A, Gilmore EA, Matias DM, Nagabhatla N, Orlov A, et al. Perspectives on tipping points in integrated models of the natural and human Earth system: Cascading effects and telecoupling. Environ Res Lett 2022;17(1):015004. doi: 10.1088/1748-9326/ac42fd.

[113] Key K. California wildfires: the holy fire at lake Elsinore on August 9, 2018. Flickr, 2018. URL: https://www.flickr.com/photos/slworking/29034137667/.

[114] Germanwatch. Global climate risk index 2021 - World. ReliefWeb, 2021. URL: https://reliefweb.int/report/world/global-climate-risk-index-2021.

[115] United Nations Environment Programme. As heatwaves blanket Europe, cities turn to nature for solutions. United Nations Environment Programme, 2022. URL: http://www.unep.org/news-and-stories/story/heatwaves-blanket-europe-cities-turn-nature-solutions.

[116] Black R. Flood risk and the UK. Energy and Climate Intelligence Unit, 2022. URL: https://eciu.net/analysis/briefings/climate-impacts/flood-risk-and-the-uk.

[117] Environment Agency and Department for Environment, Food & Rural Affairs. Prepare for flooding to reduce impacts on mental health. GOV.UK, 2020. URL: https://www.gov.uk/government/news/prepare-for-flooding-to-reduce-impacts-on-mental-health.

[118] European Banking Authority. Mapping climate risk: main findings from the EU-wide pilot exercise. Copenhagen Centre on Energy Efficiency, 2021. URL: https://c2e2.unepccc.org/kms_object/mapping-climate-risk-main-findings-from-the-eu-wide-pilot-exercise/.

[119] Geiger T, Frieler K, Levermann A. High-income does not protect against hurricane losses. Environ Res Lett 2016;11(8):084012. doi: 10.1088/1748-9326/11/8/084012.

[120] Krznaric R. The empathy effect: How empathy drives common values, social justice and environmental action. London: Friends of the Earth, 2014. URL: https://dokumen.tips/documents/the-empathy-effect-friends-of-the-earth-policy-and-insight-the-empathy-effect.html?page = 1.

[121] European Environment Agency. Air quality in Europe 2021. European Environment Agency, 2021. URL: https://www.eea.europa.eu//publications/air-quality-in-europe-2021.

[122] Wallace-Wells D. The uninhabitable Earth: A story of the future. London: Penguin Books, 2019.

[123] Raworth K. Doughnut economics: Seven ways to think like a 21st-century economist. White River Junction, VT: Chelsea Green Publishing, 2017.

[124] Myers SS, Pivor JI, Saraiva AM. The São Paulo declaration on planetary health. Lancet 2021;398(10308):1299. doi: 10.1016/S0140-6736(21)02181-4.

Chapter 4

Understanding Climate Migration with a Focus on Latin American Children and Families

Esther Tobarra[1,3,*], MD
Fernando Wang[2,3], MSc
and I Leslie Rubin[4-8], MD
[1]Cwm Taf Morgannwg University Health Board, Royal Glamorgan Hospital, Rhondda Cynon Taf, Wales, United Kingdom
[2]Oncology Institute of Marseille, Marseille, France
[3]Interuniversity Master in Planetary Health, Universitat Oberta de Catalunya, ISGlobal, Pompeu Fabra, Barcelona, Spain
[4]Department of Pediatrics, Morehouse School of Medicine, Atlanta, Georgia, United States
[5]Department of Pediatrics, Emory University School of Medicine, Atlanta, Georgia, United States
[6]Southeast Pediatrics Environmental Health Specialty Unit, Emory University, Atlanta, Georgia, United States
[7]Break the Cycle of Health Disparities Inc, Atlanta, Georgia, United States
[8]The Rubin Center for Autism and Developmental Pediatrics, Atlanta, Georgia, United States of America

Abstract

Consequences of climate change are forcing an increasing number of people around the world to migrate from their climate-stressed countries of origin to countries that offer better access to vital resources and greater

* ***Correspondence:*** Esther Tobarra, Community Paediatrician, Cwm Taf Morgannwg University Health Board, Royal Glamorgan Hospital, Rhondda Cynon Taf, Ynysmaerdy, Pontyclun CF72 8XR, Wales, United Kingdom.
Email: esther.tobarra@wales.nhs.uk.

In: Environmental Health: Leaders for Climate Change
Editors: I Leslie Rubin and Joav Merrick
ISBN: 979-8-89530-347-4

security. This trend can significantly impact migrant's health and socioeconomic outlooks, as well as having unanticipated implications for the countries of origin and destination. Many migrants are families with children, who are most vulnerable to the impacts of environmental hazards. Migration can disturb a child's formative years, increasing the risks of physical harm, mental illness, and social vulnerabilities, and, when traveling alone, becoming a target for kidnapping and human trafficking. We explore the example of an increasing number of migrants leaving countries in Latin America for neighbouring countries including the United States (US). Around half of those arriving to the US originate from the Central Dry Corridor of the Northern Triangle, with food insecurity being one of the primary reasons. There is an urgent need to protect the rights of this group of vulnerable children, ensuring their appropriate settlement, with access to healthcare and education. Health professionals can play an essential role in identifying risks to migrant children's physical and psychological health and advocating on their behalf. Social Prescribing is an example of a valuable strategy that offers a community-centred framework, favouring integration and personalised care, and taking a holistic approach to health and wellbeing. We also recognize the need for more research and information on optimal strategies to promote migrants' family wellbeing in their new country.

Introduction

The links between climate change and its long-term social and global health consequences are becoming an increasingly important topic for human rights, migration, public health, security, economics, and other professional sectors (1). Environmentally induced migration is one of the harshest outcomes of climate change involving a profound loss of resources and social capital for migrant individuals, families, and children. Moreover, migrants can be exposed to trauma and violence, with significant social, economic, and mental health and consequences, especially for children (2).

The International Organisation for Migration (IOM) (3) defined "environmental migrants" as "persons who, predominantly for reasons of sudden or progressive change in the environment that adversely affects their lives or living conditions, are obliged or chose to leave their home, either temporarily or permanently." As examples, these could be people who leave due to declining crop productivity caused by desertification or people displaced by a climate emergency such as an extreme weather event. Rising global temperatures and sea levels, increasing frequency and severity of natural disasters, and progressive depletion of life-sustaining resources are among the drivers that stimulate population mobility.

Millions of people are already migrating as a response to environmental degradation, loss of livelihoods, and adverse impacts of climate change. The World Bank estimates that by 2050 there will be 200 million people displaced worldwide due to climate change, with most movements occurring within countries (4). In 2021, 22.3 million out of the 38 million (58,6%) of total internal displacements across 141 countries were weather related - the majority of them due to storms, floods and wildfires (5). Projections forecast that current trends will rapidly accelerate. Children born in 2020 experienced a 6.8-fold increase in heatwaves, and a 2.8-fold increase in floods, compared to those born in 1960 (5).

The Children's Climate Risk Index (CCRI) formulated by UNICEF provides a comprehensive view of children's "exposure" and "vulnerability" to the impacts of climate change. It ranks countries based on "children's exposure" to climate and environmental disasters, such as cyclones and heatwaves, as well as "children's vulnerability" based on their access to essential services. This composite index aids in assessing and measuring the likelihood of climate and environmental stresses that negatively impact children's health, development and well-being, including the deepening of deprivation and humanitarian situations (6). The CCRI estimates that around a billion children, or nearly half of the world's children under the age of 18 years, live in "extremely high-risk countries". Around 820 million are exposed to heatwaves, 920 million to water scarcity and 870 million to cyclones or flooding. Heatwaves and water scarcity can also trigger drought and wildfires. Although this high level of exposure should not necessarily be equated with displacement risk, it is a strong indicator of where the risk of displacement is highest.

These exposures and children's vulnerability are also linked to poverty and to the inadequacies in city and municipal governments, the lack of infrastructure, knowledge, and capacity to adapt, which effects those who are least able to cope with the impacts, such as injury, death and loss of property and income, more significantly (7).

Children and Families' Vulnerabilities in Climate Migration

The unique concerns for children are often omitted in climate change discussions (8), yet children are disproportionately vulnerable to the impacts of climate change and migration because of physiological, developmental, behavioural, and social factors (9). The landmark 2019 Lancet countdown report on health and climate change stated that "the life of every child born today will be profoundly affected by climate change" (10). The reasons behind this are complex and multifactorial. Critical periods of development make children more susceptible to the toxins absorbed through inhalation of air pollution or ingestion of contaminated water following flooding or other food exposures related to environmental degradation and climate change (9, 11, 12).

According to World Economic Forum, half of the world's children under the age of 18 years (around a billion), live in high-risk areas for climate change, estimating that 850 million children - 1 in 3 children worldwide - live in areas with at least 4 overlapping climate and environmental hazards, including heatwaves, cyclones, floods, water scarcity, lead pollution, air pollution or infectious and vector-borne diseases. Almost every child on Earth is exposed to one of these risks, and 80 million children face at least six of these climate hazards (13) (see Figure 1).

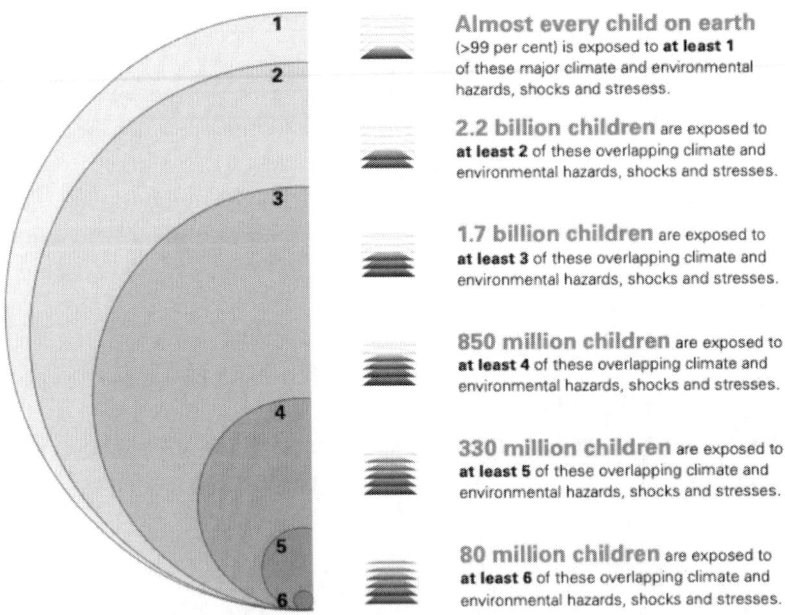

Figure 1. Number of children living in areas that experience multiple overlapping climate hazards. Image source: UNICEF (6).

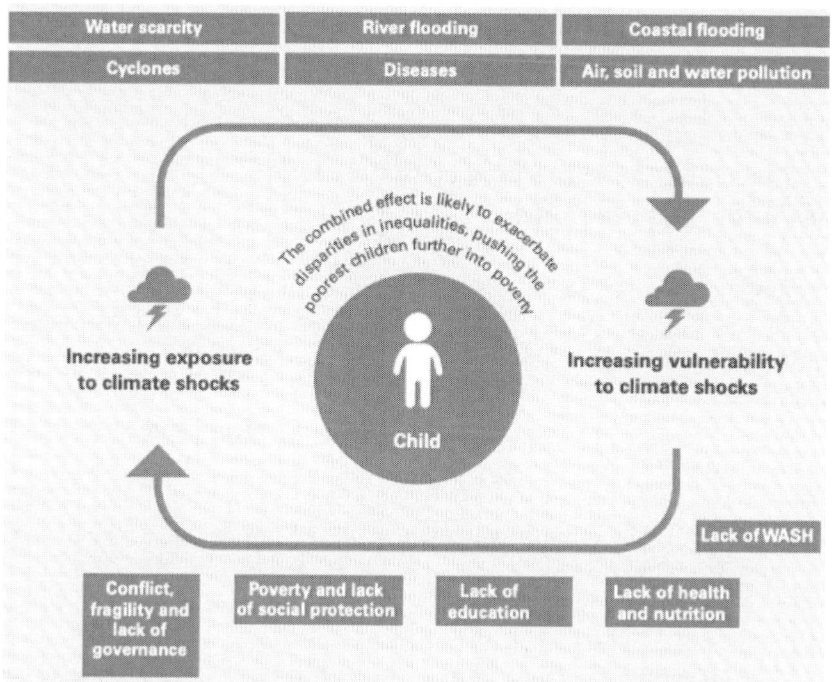

Figure 2. Children caught in a vicious cycle of increasing exposures and vulnerabilities. Image Source: Children's Climate Risk Index, UNICEF (6).

According to the Data for Children Collaborative with UNICEF, 920 million children are highly exposed to water scarcity and 820 million children to heatwaves, 400 million children are exposed to cyclones and 570 million to flooding of rivers and coasts (6). These figures are expected to increase as climate change continues to evolve. Little is known about how various multi-levelled individual, structural, cultural, institutional and geographic factors intersect to influence the lived experience of climate risk in children, but we do know that the vulnerabilities of this group increase along with the exposures and lack of water, sanitation and hygiene (WASH) in a vicious cycle (6) (see Figure 2). As a result, it can be expected that some of these complex situations trigger migration for environmental reasons.

The last Internal Displacement Monitoring Centre report, IDMC 2022, focuses on children and youth in internal displacement and highlights their 'invisibility' in displacement data (5). This has direct consequences for our collective ability to understand their needs, aspirations, and potential, and from there, to identify appropriate policy and programme responses, particularly in long-term planning to prevent and resolve displacement and assure their safety, health, and wellbeing.

Displacement and Its Impacts on the Health of Children and Families

Studies have shown strong links between displacement and self-reported health, and between poor early-life conditions associated with displacement and ill-health later in life (2). In addition, displaced children and young people are at risk of trauma by the events they have witnessed, and are at risk of neglect, abuse, and violence. Displacement can also tear families apart to the severe detriment of children's wellbeing with social repercussions in young people that can last for years and even affect future generations, due to loss of learning and future productivity, which also impacts on economies and societies as a whole (5).

Despite this, findings show that, although migrant children experience more traumatic events than non-migrants, they can also develop greater resilience than non-migrants. Further research is needed to investigate how levels of resilience may vary over time and among different countries, and how children's resilience can be promoted among migrant populations (14).

It is also important to distinguish between the different ways displacement affects children and young people, depending on their age, gender, or other factors, such as associated disabilities. For example, teenage girls are particularly at risk of exploitation, early marriage, sexual violence, and exclusion from education. Several factors increase this risk and certain groups, such as unaccompanied minors and children separated from their families, who are among the most vulnerable. Therefore, denying them appropriate attention would leave one of the most vulnerable groups of children at risk for physical illness and mental breakdown (2).

It is predictable that environmental migration has similar adverse outcomes for children and young people's health as those from general displacement, with a higher prevalence of malnutrition, childhood trauma and early pregnancy, but with added risks related to the specific environmental circumstances that triggered the migration, for example, extreme weather events and natural disasters and exposure to environmental hazards (5).

The disruption of education may eventually result in less economically productive communities, which in turn render the children less resilient to future shocks and may perpetuate a cycle of crises and health disparities for the children (see Figure 3).

Figure 3. Cycle of children's health disparities. Adapted from break the cycle of health disparities (15).

Environmental Migration in Latin America

Specialists and scholars had determined that climate migration was going to take place in areas of South-East Asia, West Africa and low-lying islands, however, the region of Latin America has a combination of environmental and non-environmental factors that converge to give rise to 'hot spots' for mass population movements (16).

Thousands of Latin Americans are currently crossing countries, many to other countries in the region, to the United States and Canada, and for those who can afford the journey, to Europe. Forced by corruption, violence and increasingly by the adverse effects of climate degradation, they leave their homes to emigrate. Unfortunately, the authorised and safe tracks are not often the first choice, with the vast majority going through secondary, unsafe, and illegal routes (17).

The Internal Displacement Monitoring Centre (IDMC), the world's leading source of data and analysis of displacement, registered 4,5 million displacements due to natural disasters in the Americas in 2020, an all-time record, added to the 1,6 million new migrations in 2021 in the region (18). The World Bank estimates that by 2050, the potential number of climate migrants will soar to 17 million in Latin America (4).

The Global Climate Risk Index (CRI) (20), an independent environmental organisation which works for sustainable global development, analyses the extent to which countries and regions have been affected by the impacts of weather-related events such as storms, floods, heat waves etc. The CRI highlights the vulnerability of the Central American Dry Corridor countries to climate impacts, with Honduras being the most affected country in the world by the consequences of environmental degradation from 1994-2013, Guatemala the 9th and El Salvador the 12th (20). In 2010, the Climate Vulnerability Monitor, (19) estimated a "severe" category for El Salvador and Honduras, the second most vulnerable category (20).

The Intergovernmental Panel on Climate Change (IPCC) views this situation as being aggravated in Latin America due to population growth and the construction of housing within high-risk areas, high rates of inequality and poverty, and the lack of sustainable planning in the exploitation of the countries' natural resources (21). Not only are the tropics – where most developing regions are located – at higher risk of experiencing natural hazards, but a combination of political, economic, and social factors lower resiliency and capacity of the populations in these regions to effectively respond to climate-related events (16). Both climate and non-climate factors such as poor governance, poverty, overpopulation, violence, organised drug-related crime, and unequal land distribution compound these migratory pressures (21).

Climate Factors for Migration in Latin America: Hurricanes and Floods, Sea Levels Rising, Drought and Melting Glaciers

Latin America is a territory with a high-risk for climate emergencies that has also shown variability in its ability to respond to natural disasters in the past, resulting in human displacements.

An iconic example of poor emergency preparedness is one of the most incomparable catastrophes that affected the poorest country in the Western hemisphere following the magnitude 7.0 earthquake that struck Haiti's capital on January 12, 2010, incurring a death toll of over 200,000 and the damage or destruction of almost 300,000 residences and commercial buildings (16). In 2021, Haiti is still considered the Latin American country with the highest CCRI, with an index of 7.3 in 29th position (6).

Other examples of well-known environmental threats that caused more recent migration and displacement in Latin America are related to El Niño, a climate phenomenon that warms ocean temperatures and has significant effects in local weather from Australia to South America (22). El Nino caused massive displacements due to severe floods in Peru in 2017, devastating fires associated with deforestation have forced the evacuation of several Amazonian communities in 2020, and Hurricane Dorian forced the displacement of 295.000 people due to floods and landslides in Brazil in 2019 (18). In 2020 Hurricane Ida triggered the most displacements, prompting nearly 14,000 pre-emptive evacuations in Cuba, causing widespread damage to homes and infrastructure, particularly to the electricity grid (5). Brazil recorded 411,000 flood displacements in 2021, prompting several municipalities to declare a state of emergency. Indigenous and rural black communities, who lost their homes and crops, struggled to meet their basic needs, including for food and potable water. Rains and floods also destroyed more than 8,000 homes in 85 municipalities across Venezuela in August 2021, triggering around 32,000 displacements (5).

As the oceans' temperature continues to rise due to climate change, tropical storms and hurricanes are expected to become increasingly more frequent and more potent. Therefore, it is reasonable to presume that as hurricanes hit the Caribbean with more frequency and ferocity, households that have repeatedly suffered from these events may increasingly consider migration as a life-saving strategy. The latter option may be facilitated by the existence of strong migration ties and networks between Latin American countries and the United States (16). Sea level rise can also have a significant effect on the Caribbean population as the majority of the islands' inhabitants live less than 1.5 km inland from the coast, and several Caribbean island-countries are at risk of being partially or completely submerged (16).

Drought is a complex and long-lasting natural hazard that is aggravated by deforestation and wildfires. Drought affects soil quality and causes erosion because of the absence of moisture, that dramatically affects agricultural potential and thus food production and livelihood. As an example, North-eastern Brazil, which is primarily an agricultural region, has experienced spikes in out-migration to the country's southern regions following decreases in crop yields related to periods of severe droughts. These intense droughts have caused a decrease in water resources and an increase in the costs of alternative sources. Around 70% of freshwater is found in the Atlantic region of the Central Dry Corridor, however the highest population densities are found in the Pacific (27). The central area of Chile is another predominantly agricultural area in Latin America that has been seriously affected by droughts (16).

Even melting glaciers have a significant effect on South America. The Andes Mountains may see a reduction in the generation of available water for agricultural and personal consumption, with an increased risk of drought and fires during the dry season, and a change in rain patterns with the risk of flooding. This can place even greater economic pressures on the poorer sectors of society, who are already sorely affected by water privatisations in the past (16).

Non-Climate Factors that Impact Migration: Social Conflict in Latin America

While environmental changes are unlikely to induce mass migrations in a country that has an accountable and responsive government, the greatest contributors to climate migration are from areas and countries where 'non-environmental' factors, such as poor governance, political persecution, and poverty, are present and already exercising migratory pressures on the local populations (16).

Some Latin American communities have shown great resilience facing cumulative environmental and socioeconomic impacts from climate and environmental degradation. However, many communities in Central America continue to live with inadequate water and sanitation as a result of poor infrastructure, poor housing and food insecurity after hurricanes Eta and Lota destroyed homes, infrastructure and livelihoods in 2020, a situation accentuated by the COVID-19 pandemic (23).

Ongoing social transformations in Latin America highlight the outrage of communities. Tensions between citizens and their governments have intensified in relation to the usufruct of land due to the liberalisation of key sectors of the Latin-American economies, such as oil and gas. Community activism is being silenced by the murder of significant number of activists in some Latin American countries. According to the Latin America Working Group, 572 journalists and activists were assassinated in Latin America and the Caribbean, from 2010 to 2015, representing 76,7% of all worldwide murders due to activism (24).

Deforestation, which is linked to illegal extractive activities most notably in the Amazonian communities, increased in the Amazon by 79% between 2019 and 2021, compared with 2016 to 2018. The rate is even higher for indigenous land, where it rose by 138% (5). Wildfires in the area are linked to deforestation in what is called "slash-and-burn cycles". Forests are cut in the wet season and burned in the dry season, often by landowners and landgrabbers for logging or with a view to converting land for agribusiness (5).

Moreover, the socioeconomic vulnerabilities of the regions translate into food insecurity for households, which are forced to considerably reduce their food intake due to market shortages. Oxfam and other humanitarian organisations have noted that from August 2019 to June 2020, the severe food insecurity situation in the Central Dry Corridor increased by 12% compared to previous periods (25). Moreover, the COVID-19 pandemic drastically reduced household income by around 20%, as well as the GDP of Guatemala, Honduras and El Salvador which has contracted between 6 and 11% (25). Food hoarding by the most privileged sectors of the countries and commercial speculation has dramatically increased the prices of basic products of daily life (25, 26). Women are especially vulnerable in this context, especially single-parent families headed by females, which represent 7 out of 10 vulnerable households (25).

There are many examples where social and political conflicts exacerbate the vulnerability of communities, and any form of adaptation loses effectiveness in the face of climatic hazards, and, as a consequence, migration emerges as the most appropriate possibility for protection of physical, mental, social and economic well-being. Therefore, individuals are forced to leave their homes in search of safer destinations where the impacts of inclement weather are better absorbed by better political planning, relatively stable infrastructure, with greater resources and financial capacities (28).

Possible Migratory Flows for Future Latin American Climate Refugees

According to the IOM, the routes through the countries of the Central American Dry Corridor, which is the area between Chiapas and Costa Rica, crossing Guatemala, Honduras, El Salvador, are now as deadly as the Central Mediterranean routes for migrants and asylum seekers from Africa trying to reach Europe. A large portion of the fatalities are Salvadorans, Guatemalans and Hondurans who cannot afford safer routes (29).

Venezuela is the country where most migrants have been forced to leave their homes as a result of internal political conditions and security problems related to drug trafficking, with a total of 5.1 million people out of 28,4 million inhabitants, representing 17,9% of its citizens (30). Interregional mobility has also been triggered by greater restrictive policies for Latin Americans in extra-regional destination countries, especially in the United States and Europe. For nationals of Andean countries (Peru, Bolivia, Ecuador), the Southern Cone (Argentina, Chile and Brazil) is their first destination (30). Since 2017, Colombia has also been a preferred destination due to the improvements in terms of peace policies in the reduction of armed conflict. In fact, according to data from July 2021, out of the 4.1 million Venezuelan refugees and migrants in the world, 43% were in Colombia (30).

The Dry Corridor of Central America: A Triangle of Climate Migration

The Central American Dry Corridor (CDC) (see Figure 4) is a territory that extends over 1,600 kilometres, from Chiapas (Mexico) to Costa Rica. It covers 30% of the Central American region and 80% of its extension runs through 3 countries mainly that make up the Northern Triangle (Corredor Seco del Triángulo del Norte, CSTN): Guatemala, Honduras and El Salvador. The CSTN has a population of 22.5 million people, of which 62% reside in urban areas and 38% in rural areas. In a 10-year period, from 2007 to 2017, immigration to the United States from the CSTN area has grown by 25%, a high percentage compared to other regions from around the world. Half of the migrants arriving in the US come from this "Triangle" although less than 25% of these actually manage to settle in the US. In 2019, almost 193,000 people were returned by the US Authorities (31).

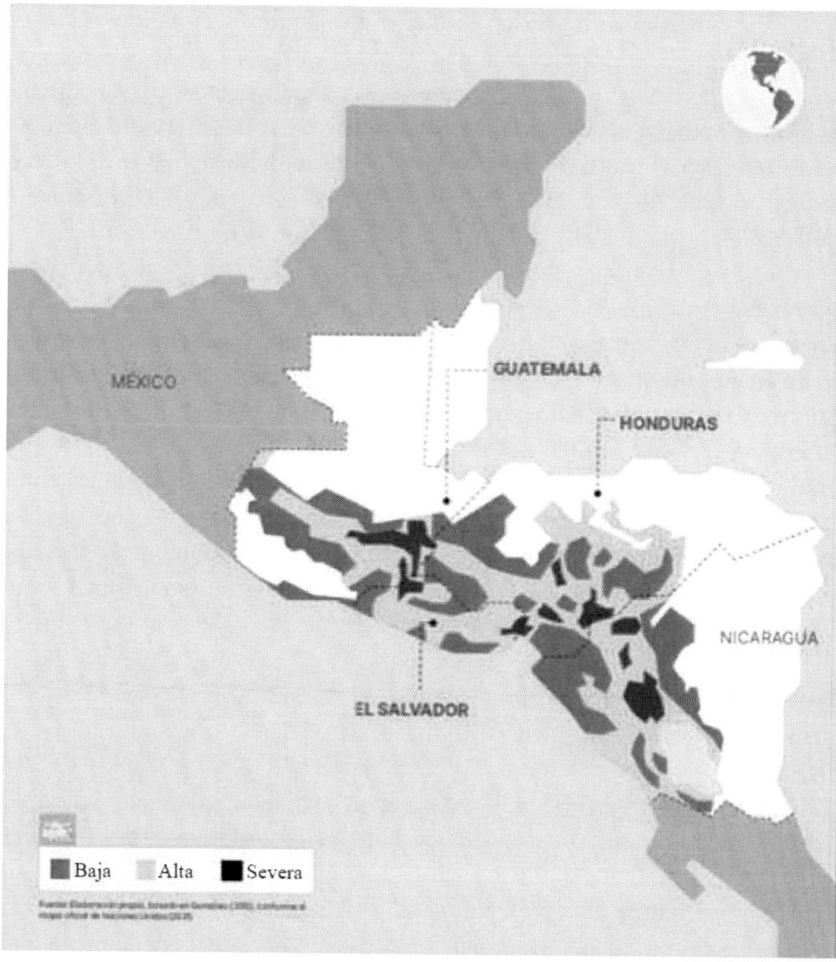

Figure 4. Central American dry corridor area map delimited by colours based on their vulnerability to climate change. Medium gray (“*Baja*”) for “low vulnerability”, light gray (“*Alta*”) for “high vulnerability” and dark gray (“*Severa*”) for “severe vulnerability”. Source: Food and Agriculture Organization (FAO.org) (32).

Flows to the United States (US)

The US is usually the main destination for many migrants looking for a new home where they hope to restart their lives. The IPCC estimates that 30 million people will be displaced from Latin America to the United States by 2050 due mainly to food insecurity because of climate crises (4).

The US border with Mexico with all its orographic peculiarities represents a potential challenge for anyone who wishes to cross. The Rio Grande, (Rio Bravo for Mexicans), is a river which navigates between the two countries. It is a large body of water in constant turbulence that has caused the death of more than 4,000 people who have dared to cross it since 2014 (33). Others decide to hide in large capacity vehicles facing extreme risks of death by suffocation or murders by criminal gangs operating in the area (33).

For many migrants, the Darién Gap, a land bridge connecting south and North America, is the only way to head north to the US (34). This lawless stretch of jungle is one of the most perilous migrant crossings in the world and people can only make the way on foot. Adult migrants and children need to survive from the threats of being washed away in raging rivers or falling down steep ravines, and run the risks of being raped, beaten or robbed by armed trafficking groups. Médecins Sans Frontières (MSF) stated that the number who made the treacherous crossing exploded in 2021 due to the economic and political impact of the pandemic (34). More than 70,000 migrants, including 13,000 children, crossed the Panamanian jungle between January and August 2021, surpassing all the figures registered the past five years in total (34).

Unaccompanied Children Arriving to the US

Minors arriving in the US without a parent or a legal guardian and without legal immigration status are called unaccompanied children (UCs). The Office of Refugee Resettlement (ORR), a program of the Administration for Children and Families within the US Department of Health and Human Services, places the children into a network of shelters (foster homes and accommodations). UCs are provided with temporary homes and services, such as health and education. The ORR states that the vast majority (92,9% of the UCs) have direct or close relatives willing to care for them. Nevertheless, more than 1,500 UCs (7.2%) had no sponsor with whom to reunite (35).

Since the beginning of the 2021 Federal Fiscal Year (FY21) in October 2020, until July 2021, 72,000 UCs arrived in the United States (35), with most UCs in ORR custody arriving from the Northern Triangle countries, El Salvador, Guatemala and Honduras. Mexican UCs represent 25% of the migrants but they do not get any state protection under the 2018 Trafficking Victims Protection Reauthorization Act (TVPRA) (35).

The proportion of migrants' gender has been stable since 2014, with 70% male vs 30% female. Age also remains stable, with the highest rates in young adults aged 15 to 16 years (40%), and 17 years of age (35%). The data shows a significant decrease in the percentages of children aged from 0 to 5 years of age in FY21, representing only 1,8% of the total arrivals in May 2021 (35).

Young adults aged between 16-18 years of age are less likely to be placed in foster homes and more likely to be placed in temporary accommodations. Most often they do not have the right to work or are not enrolled in any full-time education. The lack of sufficient arrangements and provision for this group makes them one of the most excluded in society. Many authorities of different countries do not offer this age group a full needs assessment (36).

Strategies and Innovative Practices to Address the Needs of Climate Refugees Within the Health Sector

The 2020 Lancet Countdown Report on Health and Climate Change notes that "Doctors, nurses, and the broader health profession have a central role in health system adaptation and mitigation, in understanding and maximising the health benefits of any intervention, and in communicating the need for an accelerated response" (10). Leading experts urge health professionals to be leaders in addressing climate change in clinical practice, meaning disseminating knowledge to individuals and communities and serving as advocates for policies to improve the health of populations (37, 38).

Adaptation Strategies: A Call to Preventive Aaction

Governments need to recognize the need for urgent actions to adapt to climate change. There are many examples of unacceptable delays and insufficient responses to climate-related emergencies in Latin America, and there are many strategies that can enable countries to adapt and respond better to impacts of climate change. For example, retrofitting buildings and hospitals by making them weather-resistant will ensure continuation of care that is dependent on electrically powered medical equipment (e.g., ventilators) and prevent flood-related hospital evacuations in times of danger. To the extent that hospitals can anticipate these dangers and adapt ahead of time, they will be better able to serve their communities during weather-related disasters when access to urgent care is most important. In this sense, successful case studies can lead the way, for example:

After conducting an assessment that showed the building's vulnerability to wind damage, the Nicklaus Children's Hospital in Miami secured funding through FEMA's Hazard Mitigation Grant Program (HMGP) and looked for a way to improve its safety with a Hurricane-Resistant Shell made of panels of glass, fibre and reinforced concrete that serves as a protective cocoon around the hospital. It was designed to withstand winds of up to 200 miles per hour and proved successful, as during hurricane Francis in 2004, none of the young patients and their families needed to evacuate, and the hospital served as a temporary refuge for nearly 1,000 staff members and their families, ensuring staff recruitment. The adaptation strategy has a long life preventing further damage proved with Hurricanes Katrina and Wilma in 2005 (39).

In contrast, the situation of disaster prevention and relief in Central America, where hundreds of thousands of people are periodically left homeless during the hurricane season, may serve as an illustration of the low capacity of response, and highlights a gap that needs urgent action.

Right to Health

Politicians and lawmakers around the world are called upon to act immediately to improve environmental policies with greater forcefulness to face the consequences of climate change. In November 2021, US President Joseph Biden released the first White House "Report on the impact of climate change on migration" (40), which officially recognized a link between climate change and migration. However, current legal instruments still do not protect individuals displaced by climate change impacts in the US. Politicians continue to debate whether they should be called "climate migrants", "internally displaced persons" or "refugees". Policymakers argue that there should be a distinction between those displaced by one-time events and those displaced because of no-return events such as rising sea levels.

Asylum seekers and refugees are at increased risk of physical and psychological health problems. Across the world, there are well-established assessments of the physical health needs of refugees on arrival, but mental health and wellbeing have received less attention and are one of the main long-term primary health concerns in this group, so efforts to address their psychosocial and mental health needs are worthwhile (41).

Children, as environmental migrants, need effective representation and dedication as they have little or no agency in decision-making with regards to the life-changing extreme scenarios that they and their family need to endure. It is estimated that around 40% of refugees or asylum seekers across the world are children; and the continuity of their education, everyday activities, access to stable environments and routines, as well as accessibility to healthcare services, is severely compromised as a result of extreme situations that are out of their control (5). It is fundamental to envision and ensure Children's Rights when aiming to mitigate and deal with children's climate risk (6). National legislation in some countries can serve as illustrations that lead the way, where at least child migrants' equitable 'right to health' is formally recognised, although there remain challenges and barriers to care that impede translation into practice (42).

According to a literature review on the benefits from preventive mental health interventions for refugee children and adolescents in high-income settings, a continuum of care with multi-level and cross-sectoral intervention models is needed to address the multitude of acculturative and resettlement stressors faced by resettled refugees' families (43). This review concludes that a focus on parents, including their mental health needs and how stressors are managed within the family including variance in assimilation for different family members, might improve mental health outcomes for children. Also, school-based interventions hold promise as access to refugee children might be easier within the school and interventions to support peer relationships and a sense of belonging are important to consider.

Appropriately addressing the needs of these children and families will result in increased benefits, not only to the children as individuals, but to the whole society in the country of destination, by preventing ill health and promoting wellbeing resulting in greater opportunities for employment and lower risk of crime. More data is needed to better understand the broader and longer-term impacts, but it is clear that protecting and supporting displaced children and young people not only safeguards their rights, but also contributes to a more stable future for all. Thus, addressing their needs in a timely and appropriate manner has the potential to reduce pressure on urgent healthcare and avoid escalation and greater need for resources later on (5).

Social Prescribing

Asylum seekers and refugees often experience poor access to healthcare in host countries as they navigate unfamiliar health systems. Results of a qualitative study in the United Kingdom (UK) about refugees' experiences in accessing healthcare, identified the main barriers as relating to language difficulties, health literacy, unrecognised needs, and the cost of travel to appointments. Some feared seeking support for mental health from their General Practitioners and few participants were aware of health prevention services (44).

Therefore, it is unreasonable to expect refugees to fit into existing patterns of healthcare delivery, and they require substantial support in accessing services and adapting to a new country. Challenges include knowledge about entitlements, availability of professional interpreters, and access to specialist and preventive services.

The complex health needs of refugees benefit from a multi-agency or a holistic, family-orientated approach, tackling wider determinants of health. Ideal interventions need to be culturally informed and sensitive to their background, competences that are usually outside the experience of doctors and other health professionals, teachers, and social workers, who would require guidance on how to respond effectively to the specific needs of refugees, and more so for climate refugees (45).

Social prescribing, also known as "community referral", is a concept that has gained support in the National Health Service (NHS) of the United Kingdom as well as in Ireland and the Netherlands and forms part of the NHS long-term plan, since a Commission in the UK estimated that about 20% of patient consultations were for social problems rather than medical problems (36). Social prescribing is defined as a method by which referrals from healthcare professionals, statutory services, voluntary services as well as self-referrals are made to "link workers" or social prescribers, who in turn connect individuals with community, voluntary, statutory, and other sector services to improve personalised and holistic health and wellbeing (46). The prescribed activities to which link workers refer individuals are often non-clinical in nature, and may include physical activity groups, as well as arts and crafts, befriending, education and training, volunteering opportunities, and health-related therapies. A review study showed that social prescribers require appropriate training on how to support migrants to address the wider determinants of health and in this case, they might take on additional support roles and/or actively deliver parts of the prescribed activities themselves, which is outside of the scope of the typical link worker role (45).

Moreover, findings from needs assessments done for families of refugees highlighted the need for accessible foundation-level mental health promotion resources, to encourage self-help and mental well-being, and early access to adult primary mental health services or Child and Adolescent Mental Health Services (39). Broader contextual factors such as language barriers, housing and living conditions, integration in the community and being able to work and contribute to social and economic capital can potentially improve mental health outcomes for migrants.

In this sense, by co-designing a personalised intervention there is greater likelihood that the outcome will be useful and beneficial. There are a few initiatives related to displacement that involve children and young people in problem analysis and solutions planning that have been found successful (5). Examples of good practices exist, such as the Philippines' Children's Emergency Relief and Protection Act. Developed in consultation with young people, it is the first national policy in the world to set out initiatives both to prevent displacement and protect displaced children (47). It includes a range of child-centred measures to meet basic needs, establish evacuation and transitional centres for orphaned or unaccompanied children and increase protection against child trafficking and abuse. It also focuses on measures to improve data collection and monitoring (5). The Honduran organisation "Jóvenes contra la Violencia," or Youth against Violence, has been recognised for its community-based work with internally displaced people. It also successfully presented a bill on displacement for debate in the national congress (48). This and many other examples reveal young people's willingness to engage as first responders in crises and key figures in community reconstruction and peacebuilding after disasters, conflict and displacement (5).

Additionally, although social prescribing is an innovative area for intervention in this group and there is a lack of specific research to demonstrate clearly definable outcomes, improved self-esteem, confidence, empowerment, and social connectivity were frequently described by migrants as a result of Social Prescribing support (45). A social prescribing framework could address different climate refugees' needs. A social prescriber could play the following roles:

- Be a multiagency link worker: allocated at arrival (crucial stressful time): Help families to navigate between systems: Social Services, Health, Home Office, Education, Legal Advisers.
- Recognise wider determinants of their health and be culturally informed/matched. Establish culturally acceptable and not stigmatising interactions.
- Understand their complex needs and develop a personalised tailored care and support plan – with greatest value of "what really matters to them"
- Advocate and safeguard their vulnerabilities
- Help the family to find environments that feel comfortable and supportive.
- Introduce community groups and support services – activities that do not rely on language. It promotes integration, grows confidence and builds a network of support
- Promote independence – Coaching and information about how to get skills or qualifications.

To improve the sustainability and efficiency of a Social Prescribing model for migrants, it is worthwhile to offer migrants who are already established in the country, opportunities to become Social Prescribers as they will have first-hand lived experience that will be valuable in helping others arriving in the country.

Potential Positive Consequences of Migration for Economies and Societies

How displacement of populations has an effect on host societies is complex and our knowledge and understanding is evolving. In this section we will focus on the integration of Latin American (Latinx) populations in the US society. Latin American migration has been and continues to be invaluable to the US by its important and valuable contribution in the economy as a workforce, not only by taking jobs that are difficult to fill, but also as investors and consumers, and due to their talents and new ideas that enrich cultural exchange (49).

The Latinx population can be seen as a valuable resource for the US. Latinx households hold considerable earning and consumer power. In 2017,

Latinx households earned more than $1 trillion which allowed them to pay more than $252.2 billion in taxes, including $165.9 billion in federal income taxes, as well as $86.3 billion in state and local taxes. Latinx families held more than $780.7 billion in spending power after taxes (49).

US institutions are increasingly affected by and dependent on the Latinx population. For example, the business community is increasingly relying on Latinxs as entrepreneurs, investors, and consumers. The bilingual and bicultural nature of the Latinx population also provides a valuable resource as the US business community expands its consumer markets and business operations into Latin America (50).

The higher education system is increasingly finding the Latinx populations among the ranks of students and educators. Political institutions find that Latinx citizens play a powerful role in the outcome of elections, both as voters and as political candidates. Additionally, the health care system sees Latinx populations as health care recipients and providers. Religious institutions find that their potential adherents and leaders are increasingly Latinx (50).

These trends are well underway in the largest states and at the national level, with the rising dispersion of Latinx population groups into parts of the country that traditionally have not had Latinx populations, suggests that all parts of the country will, in time, feel the impact of the growth of the Latinx population (50). These changes will inevitably require multifaceted local, regional and national strategies to facilitate and maximise the benefit for all.

Conclusion

The manifestations of climate change are already having an increasingly powerful impact on South-North human migration. In this paper, we highlight the particular situation of the American continent, with well-established migration channels between most Latin American countries and the US.

Since 2009, United Nations University (UNU) report *In search of shelter* (51) predicted that the loss of 'ecosystem services' such as arable soil, clean air, and water, as a result of the intersection of climate and non-climate factors, would be the principal cause of mass environmental migration. In a region such as Latin America where, in addition to current economic forces, historical and political factors have fatefully contributed to these displacements, it becomes critically important to analyse and include Latin America in the future debate of environmentally prompted climate migration (16).

As a result of environmental migration, families seeking asylum and undocumented environmental migrants, especially child climate refugees, face significant challenges. They may present with vulnerabilities and several potential health needs, but they are also a resilient group with the capability and determination to survive and to flourish if their complex physical mental and social needs are addressed (14).

Developed nations such as the US hold significant responsibility vis-à-vis the anthropogenic climate change that their industrial activities helped engender (16), as well as their economic statecraft influencing the trajectories of Latin American and Caribbean countries. However, the present US migration policy does not give sufficient consideration to environmental factors. Experts from the UN Human Rights Council argue that climate refugees need access to the same protected status offered to refugees that have escaped conflict, which includes the right to accessible health services.

Doctors, nurses, and the broader health profession community have a central role in climate change adaptation and mitigation, as well as communication and advocacy for climate refugees. Furthermore, it is estimated that the healthcare sector alone contributes up to 10% of total US greenhouse gas emissions (37). As health care professionals, preventing and mitigating the health problems attributed to climate change may be our biggest challenge yet.

In this article we offer some strategies that could potentially break "the cycle of health disparities" for climate migrants, with innovative practices that aim to address refugees' social determinants of health such as social prescribing. There is evidence for preventive interventions aiming to meet the complex needs of this group, to help families to navigate unfamiliar systems (housing, legal issues, family reunion entitlements) at arrival, and to tackle their isolation and wellbeing needs, promoting their confidence, independence and integration into society, where their contribution will be valuable (41, 43).

Finally, it is known that children are disproportionately more affected by climate change, but their rights are less visible and often neglected in the complex scenario of climate change migration. There remains a notable absence of children's agency and decision-making capacities within the existing research. However, children offer unique opportunities for a brighter future because of their resilience, adaptability, and potential to become agents of change (5).

In conclusion, it is urgent to support displaced families and children not only to protect their immediate integrity and rights, but also to promote their health and well-being as a long-term contribution to sustainable development and equitable societies.

Acknowledgments

There was no funding allocated to this work, however the authors would like to thank the Break the Cycle of Health Disparities Inc, as well as the PEHSU network, specifically Region 4 PEHSU at Emory University for its training and constant work in promoting paediatric environmental health. We would also like to express gratitude to the interuniversity Master of Planetary Health with Open University of Catalunya, Universidad Pompeu Fabra and ISGlobal Institute, for providing not only academic resources, methods and formal education about current environmental issues, but also opportunities for networking among its students, making possible collaborative projects such as www.redsaludplanetaria.com. This is a website that aims to convey knowledge and understanding about Planetary Health topics and pursues to promote the creation of a Spanish-speaking community of like-minded professionals interested in Planetary Health. Esther Tobarra and Fernando Wang contributed equally to this article.

References

[1] Devlin MK, Grey MA. Climate change refugees and public health implications. In: Leal Filho W, Wall T, Azul AM, Brandli L, Özuyar PG, eds. Good health and well-being. Encyclopaedia of the UN sustainable development goals. Cham: Springer, 2020:91-9.

[2] Shultz JM, Rechkemmer A, Rai A, McManus KT. Public health and mental health implications of environmentally induced forced migration. Disaster Med Public Health Prep 2019;13(2):116–22.

[3] International Organization for Migration. MC/INF/288. Ninety-fourth session. Discussion note: Migration and environment. IOM 2007:1–8. URL: MC/INF/288 - Discussion Note: Migration and the Environment (2007), International Organization for Migration (iom.int).

[4] Rigaud KK, Sherbinin A, Jones B, Bergmann J, Clement V, Ober K, et al. Groundswell report - preparing for internal climate migration. World Bank Group 2018. URL: https://openknowledge.worldbank.org/handle/10986/29461.

[5] International Displacement Monitoring Centre. Global report on internal displacement 2022. Part 2: children and youth in internal displacement. IDMC 2022. URL: https://www.internal-displacement.org/global-report/grid2022/.

[6] United Nations Children's Fund (UNICEF). The climate crisis is a child rights crisis: Introducing the children's climate risk index. Race to zero. UNICEF 2021. URL: https://www.unicef.org/reports/climate-crisis-child-rights-crisis.

[7] Hardoy J, Pandiella G. Urban poverty and vulnerability to climate change in Latin America. Environ Urban 2009;21(1):153515452.

[8] Currie J, Deschênes O. Children and climate change: Introducing the issue. JSTOR 2016;26(1):3-9.

[9] Leffers JM. Climate change and health of children: our borrowed future. J Pediatr Health Care 2022;36(1):12-19.

[10] Watts N, Amann M, Arnell N, Ayeb-Karlsson S, Beagley J, Belesova K, et al. The 2020 report of the Lancet countdown on health and climate change: Responding to converging crises. Lancet 2021;397(10269):129-70.

[11] Bevacqua J. Manufactured environmental toxins and children's health: an evidence-based review and anticipatory guidance. J Pediatr Health Care 2013;27:13-22.

[12] Etzel R, Balk S, eds. Pediatric environmental health, 4th ed. Itasca, IL: American Academy Pediatrics, 2018.

[13] Masterson V. Climate change: half the world's children are in high-risk areas. World Economic Forum 2021. URL: https://www.weforum.org/agenda/2021/09/unicef-children-climate-risk-index/.

[14] Gatt JM, Alexander R, Emond A, Foster K, Hadfield K, Mason-Jones A, et al. Trauma,

[15] resilience, and mental health in migrant and non-migrant youth: An international cross-sectional study across six countries. Front Psychiatry 2020;10:997.

[16] Break the Cycle of health disparities. URL: https://www.breakthecycle program.org/.

[17] COHA. Climate migration in Latin America: A future 'flood of refugees' to the north? The Council on Hemispheric Affairs 2010 Feb 22. URL: https://www. coha.org/climate-migration-in-latin-america-part-1/.

[18] International Displacement Monitoring Centre. Global report on internal displacement 2021: Internal displacement in a changing climate. IDMC 2021. URL: https://www.internal-displacement.org/global-report/grid2021/.

[19] Pisani M. Migrantes ambientales: el rostro humano del cambio climático en América del sur. El País 2022. URL: https://elpais.com/planeta-futuro/red-de-expertos/2022-07-19/migrantes-ambientales-el-rostro-humano-del-cambio-climatico-en-america-del-sur.html.

[20] DARA Climate vulnerability monitor. Daraint.org 2010. URL: https://daraint.org/ climate-vulnerability-monitor/.

[21] Lennox J, Ramirez D, Oliveras J. Cambio climático en centroamérica: Impactos potenciales y opciones de política pública. Naciones Unidas 2015. URL: http://estadisticas.sieca.int/documentos/ver/CambioClimáticoenCA_ImpactosPoten ciales.pdf.

[22] Castellanos E, Lemos MF, Astigarraga L, Cuvi N, Huggel C, Chacon N, et al. Central and South America. In: Climate change 2022: Impacts, adaptation and vulnerability contribution of working group II to the sixth assessment report of the

intergovernmental panel on climate change. CUP 2022:1689–816. doi:10.1017/9781009325844.014.

[23] NOAA. What are el Nino and la Nina? National oceanic and atmospheric administration website service 2021. URL: https://oceanservice.noaa.gov/facts/ninonina.html.

[24] Barrantes SA, Langman MV. Communities affected by hurricanes Eta and Iota are threatened by food insecurity, displacement and the climate crisis. International Federation of Red Cross and Red Crescent Societies 2021. URL https://www.ifrc.org/press-release/communities-affected-hurricanes-eta-and-iota-are-threatened-food-insecurity.

[25] Sánchez-R M, Riosmena F. Cambio climático global, ecología política y migración. Revista Estudios Soc 2021;(76):2-6.

[26] Equipo Regional Humanitario de Oxfam en América Latina y el Caribe. Aquí lo que hay es hambre: hambre y pandemia en centroamérica y venezuela. Oxfam Internacional 2020. URL: https://lac.oxfam.org/latest/policy-paper/aqu%C3%AD-lo-que-hay-es-hambre-hambre-y-pandemia-en-centroamérica-y-venezuela.

[27] BBC News Mundo. El corredor seco de centroamérica, donde millones de personas están al borde del hambre y la pobreza extrema por el coronavirus y los desastres naturales. BBC 2021. URL: https://www.bbc.com/mundo/noticias-america-latina-56407243.

[28] Banco Interamericano de Desarrollo, CEPAL. Proceso regional de las Américas: foro mundial del agua 2018: informe regional América Latina y el Caribe: resumen ejecutivo. 2018. doi: http://dx.doi.org/10.18235/0001028.

[29] Foresight. The future of food and farming. Executive summary. The government office for science 2011. URL: https://assets.publishing.service.gov.uk/government/uploads/system/uploads/attachment_data/file/288088/11-547-future-of-food-and-farming-summary.pdf.

[30] Noticias ONU. Las rutas de la migración en América, cada vez más mortales. Naciones Unidas 2019. URL: https://news.un.org/es/story/2019/11/1465471.

[31] Datos migratorios en América del Sur. Portal de datos mundiales sobre la migración 2021. URL: https://www.migrationdataportal.org/es/regional-data-overview/datos-migratorios-en-america-del-sur#:~:text=A%20julio%20de%202021%20se,455.500%20(R4V%2C%202021).

[32] Solera E. Caracterización del corredor seco del triángulo norte de centroamérica. Hábitat para la Humanidad Internacional 2021. URL: https://www.habitat.org/sites/default/files/documents/Resumen-CorredorSeco-ESP-8Abr22.pdf.

[33] Seco C, Salvador E. Franja de oportunidades. FAO 2022. URL: https://www.fao.org/3/cb5228es/cb5228es.pdf.

[34] Reilly P. Migrant death toll up to 9 in deadly Rio Grande crossing attempts. New York Post 2022. URL: https://nypost.com/2022/09/03/migrant-death-toll-up-to-9-in-deadly-rio-grande-crossing-attempts/.

[35] Taylor L. The growing health crisis on the world's most perilous migrant crossing. BMJ 2022;376:o419.

[36] USCRI Factsheet: Arriving unaccompanied children. Refugees.org 2021. URL: https://refugees.org/wp-content/uploads/2021/04/USCRI-UC-Factsheet-Apr-21.pdf.

[37] Office for Health Improvement and Disparities. Social prescribing: Migrant health guide. GOV.UK 2021. URL: https://www.gov.uk/guidance/social-prescribing-migrant-health-guide.

[38] Swartz MK. Taking on climate change through a health care lens. J Pediatr Health Care 2019;33:623.

[39] Ahdoot S, Pacheco SE. Global climate change and children's health. Pediatrics 2015;136:e1468-84.

[40] US Climate Resilience Toolkit. Retrofitting a children's hospital with a hurricane-resistant shell. toolkit.climate.gov 2022. URL: https://toolkit.climate.gov/case-studies/retrofitting-childrens-hospital-hurricane-resistant-shell.

[41] The White House. Report on the impact of climate change on migration. whitehouse.gov 2021. URL: https://www.whitehouse.gov/wp-content/uploads/2021/10/Report-on-the-Impact-of-Climate-Change-on-Migration.pdf.

[42] Khanom A, Alanazy W, Ellis L, Evans BA, Fagan L, Glendenning A, et al. The health experiences of asylum seekers and refugees in Wales: Technical report of the HEAR study. Public Health Wales NHS Trust, Swansea University 2019. URL: http://cronfa.swan.ac.uk/Record/cronfa50916.

[43] British Medical Association. BMA refugee and asylum seeker health resources. British Medical Association BMA House 2019. URL: https://www.bma.org.uk/media/1838/bma-refugee-and-asylum-seeker-health-resource-june-19.pdf.

[44] Fazel M, Betancourt TS. Preventive mental health interventions for refugee children and adolescents in high-income settings. Lancet Child Adolesc Health 2018;2:121-32.

[45] Khanom A, Alanazy W, Couzens L, Evans BA, Fagan L, Fogarty R, et al. Asylum seekers' and refugees' experiences of accessing health care: a qualitative study. BJGP Open 2021;69(Suppl 1):1-11.

[46] Zhang CX, Wurie F, Browne A, Haworth S, Burns R, Aldridge R, et al. Social prescribing for migrants in the United Kingdom: A systematic review and call for evidence. J Migr Health 2021;4:100067.

[47] NHS. Social prescribing. URL: https://www.england.nhs.uk/personalisedcare/social-prescribing/.

[48] UNICEF. No place to call home. Protecting children's rights when the changing climate forces them to flee 2017. URL: https://www.unicef.org.uk/wp-content/uploads/2017/04/No-Place-To-Call-Home.pdf.

[49] UNHCR. Honduran wins the 2021 regional UNHCR Nansen refugee award for the Americas, acnur.org 2021. URL: https://www.acnur.org/noticias/ul/2021/9/6155d3cd4/honduran-wins-the-2021-regional-unhcr-nansen-refugee-award-for-the-americas.html.

[50] New American Economy Research Fund. Power of the purse: the contributions of Hispanic Americans. research.newamericaneconomy.org 2020. URL: https://research.newamericaneconomy.org/report/hispanic-americans-2019/.

[51] Saenz R. Latinos and the changing face of America. Population Reference Bureau 2004. URL: https://www.prb.org/resources/latinos-and-the-changing-face-of-america/.

[52] Warner K, Ehrhart C, De Sherbinin A, Adamo S, Chai-Onn T. In search of shelter: mapping the effects of climate change on human migration. Boekenplan 2009. URL: https://www.ciesin.columbia.edu/documents/clim-migr-report-june09_final.pdf.

Chapter 5

Implications of Climate Change on Children's Health: An Equity Perspective

Martha Berger[1,*], MPA
Nse Obot Witherspoon[2], MPH
Justin Harris[1]
and I Leslie Rubin[3-7], MD

[1]Climate Equity Collaborative, Break the Cycle of Health Disparities Inc, Atlanta, George, United States of America
[2]Children's Environmental Health Network, Washington DC, United States of America
[3]Department of Pediatrics, Morehouse School of Medicine, Atlanta, Georgia, United States of America
[4]Department of Pediatrics, Emory University School of Medicine, Atlanta, Georgia, United States of America
[5]Southeast Pediatric Environmental Health Specialty Unit, Emory University, Atlanta, Georgia, United States of America
[6]Break the Cycle of Health Disparities Inc, Atlanta, Georgia, United States of America
[7]The Rubin Center for Autism and Developmental Pediatrics, Atlanta, Georgia, United States of America

Abstract

Climate change is a significant threat to human health as greenhouse gases accumulate, temperatures rise, sea levels rise, and extreme weather events become more intense and frequent. In communities throughout the

---

[*] ***Correspondence:*** Martha Berger, Director for Children's Health, Climate Equity Collaborative, VP Outreach, Break the Cycle of Environmental Health Disparities Inc, 750 Hammond Drive, Building 1, Suite 100, Atlanta, GA 30328, United States.
Email: Martha.climatecollaborative@gmail.com.

In: Environmental Health: Leaders for Climate Change
Editors: I Leslie Rubin and Joav Merrick
ISBN: 979-8-89530-347-4

world, those most often and most severely affected by climate change are those with the least means to protect themselves and to recover when disaster strikes, most notably children. The physical and mental health implications of climate change for children, from before birth, through infancy, during early childhood, and as teenagers, are distinct in the short- and long-term health effects of environmental exposures. For children in under-resourced communities, where adverse environmental exposures are more likely, and the health effects are disproportionately more harmful. Furthermore, the current trajectory of climate change is creating both anxiety and positive energy in the youth of today to become part of advocating for and making positive changes to address climate change and improve their physical and mental health. In addition, concerned citizens are forging new alliances and partnerships to address the impacts of climate change on children's health. There is an urgent need to increase awareness and knowledge among health care professionals and the general public about climate change and its disproportionate impact on children, particularly on children in poverty and marginalized communities, and to develop collaborative strategies to combat climate change, protect the children and their future.

Introduction

Climate change is a human health concern and the need to take meaningful protective action is urgent. The effects of climate change on health are not the same for all members of society (1). Children are more vulnerable to the health effects of climate change in ways that adults are not, they are exposed in unique ways, and they are uniquely sensitive to environmental threats in ways that adults are protected, and they have fewer adaptive capacities than adults (2). Studies of environmental exposures in early life confirm that the effects of climate change on children and future generations will be enduring (3). While childhood is a uniquely vulnerable life stage for experiencing the health effects of climate change, not all children are affected to the same degree (4). Children in marginalized or neglected communities suffer disproportionately from cumulative exposures and limited supportive resources (5). Yet understanding, communicating, and acting on the children's health dimensions of climate change, particularly from disadvantaged communities, are too often overlooked.

Climate Change, Health and Equity

Climate change exposes people and communities to extreme heat events, increasing air pollution, wildfires, water scarcity or overabundance, storms, infectious diseases, allergic illnesses, overstressed energy infrastructure, food, housing and transportation system shortfalls, increased conflict, and population displacement (6). The effects of these powerful environmental forces on human health can be seen as direct and indirect, acute and chronic, physical and emotional, that result in significant morbidity and mortality. Many factors influence individual or community exposure and vulnerability to the human health outcomes of climate related events, including geography, age, income, race, education, occupation, housing status and baseline health.

Vulnerabilities to climate change vary across society, with some segments of society suffering disproportionate exposures and more adverse health outcomes. Exploration into these realities provides helpful insights for discussions about climate priorities, adaptations, and mitigation. Furthermore, representation matters in discussions on climate change and equity, and voices that center on specific and identifiable vulnerable populations all critical for increasing understanding and taking collaborative and health protective action. These voices and representation should include those living in the global south, small island nations, poor people, women, farmers and outdoor workers, minorities, those affected by structural racism, those with little education, those with pre-existing medical conditions, persons who are homeless, persons who are disabled, migrant, immigrant, and indigenous groups, the LGBTQ community, the elderly, and children. As a global community, it is our responsibility to consider, understand and act on the specific needs of all members of society as consideration of these diverse needs necessary to develop a comprehensive and inclusive set of strategies to protect the health and wellbeing of all humanity and ecosystems on which we are all dependent (7).

Environmental Health: Childhood is Different

Representing 30% of the world's population, and a two-decade life stage of critical importance for development and preparation for future endeavors, children experience and interact with their environment and experiences in dynamic ways that affect their health, positively and negatively. From before birth, in infancy, in early childhood, through school age years, and as teenagers and adolescents, each stage of childhood has unique environmental exposures and experiences that influence their health, development, growth, learning, and ability to thrive in adulthood. The field of children's environmental health is the study of these unique exposures, their health effects, and the most effective ways to reduce and eliminate harmful exposures and promote positive health, growth, and development (8).

Because children's organs and organ systems are still in the process of developing, they are more vulnerable than adults to environmental pollutants in their air, water, soil and food. Children breathe more air more quickly than adults, they drink more water and eat more food per pound of body weight than adults, thus potentially being exposed to higher levels of pollutants in air, water, and soil. Furthermore, their immature metabolic systems are not able to degrade and eliminate the chemicals, which can then cause greater harm. Common behaviors of childhood, such as crawling on the floor, playing in the dirt, frequent hand to mouth activity, and spending more time playing outdoors, increase the likelihood of exposure to potentially harmful chemicals in the air, water and soil, both indoors and outdoors. Harmful exposures may also come from potentially toxic chemicals in toys and consumer goods, in food, in homes, in cars, and at childcare centers, schools, and in playgrounds. Toxic exposures during childhood can persist into adulthood, with lifelong adverse consequences on health, wellbeing, and productivity (2).

While all children are more vulnerable to adverse environmental exposures and experiences, some children have more harmful exposures and experiences than others, with greater likelihood of consequent health impairments. In the United States, children born into or living in poverty, particularly children of color, experience disproportionate harmful exposures with consequent adverse and health effects when compared to children from more affluent communities (4). Globally, poverty is a powerful ubiquitous and ever-present threat to child health and welfare (3).

Childhood in the Changing Climate

Representing about 33% of the global population, today's children did not contribute to the climate change problem, and yet will experience it as a lifelong reality. Too many children, in communities all over the world, will feel the effects of climate change as an added burden, a multiplier effect, that complicates and further burdens already difficult living conditions (7).

Children are uniquely vulnerable to climate change and climate disasters, with limited adaptive capacity. While climate change compounds and amplifies environmental threats for all children, these negative environmental forces are exacerbated for poor children and children who are marginalized by society for any reason, notably, race, faith, citizenship status, or ethnicity. UNICEF considers almost all children at risk from climate change, with almost half of all children at extremely high risk, with about 850 million children being exposed to four or more climate stressors and 820 million children currently highly exposed to heatwaves (7). Minority and marginalized youth currently experience the disparity of effects from climate change (7).

Children have unique and adverse health effects from exposure to extreme heat, poor ambient or indoor air quality, wildfire smoke, flooding, storms, vector-borne infections, poor water quality, insufficient or overabundant water quantity, and changes to the food supply in both quality and quantity. These climate change exposures may be related to a singular event, such as a flood, or they may be combined, as when a heat wave happens amid a wildfire event. Children suffer both physical and mental health consequences from these exposures (7, 9).

Exposure by Age and Stage of Childhood

To best consider the full range of childhood health effects, it may be helpful to examine exposures and effects by life stage: on the developing fetus, newborns, infants, toddlers and preschool children, elementary school age children, and teenagers. Age stratification captures the unique exposure pathways, developmental needs, behavioral patterns, and health implications of climate effects. While explicit consideration of childhood is often neglected, more so is the appreciation of childhood life stages. Research has revealed some critical vulnerabilities related to childhood life stages and the health effects of climate change.

Prenatal exposure to extreme heat is associated with maternal preeclampsia, lower birth weight or pre-term birth, as well as congenital heart defects. Similarly, prenatal exposure to particulate matter air pollution may lower birth weight, lead to preterm birth, and is associated with increased infant mortality (10). Prenatal exposure to wildfire smoke may affect birthweight and preterm delivery, with the timing of exposure, by trimester, playing a role (11). Maternal exposure to PM 2.5 in the second trimester of pregnancy, may be associated with increased blood pressure in childhood (12)

Infants are not able to regulate their body temperature and are reliant on adults to take care of them, to keep them from overheating and to keep them hydrated. Infants cannot communicate thirst or overheating clearly and have no agency to escape heat on their own. While studies often group health effects on infants with young children, findings show that infants under one year of age are particularly vulnerable to heat-related deaths (13).

Young children are also reliant on adults to take care of them and are also vulnerable to excess heat, and behaviorally they are less likely to be aware of thirst or to stop active play to drink water as temperatures rise. Young children exposed to wildfire smoke are more likely to have respiratory health effects, with respiratory infections, and increase in ER visits for asthma (12).

School-aged children, from kindergarten through middle school, tend to spend more time in physical activity out of doors, with attendant exposure to heat and poor ambient air quality. In disasters such as floods, children this age may be at greater risk of ingesting polluted flood waters (14).

Activities and patterns of behavior put older children at risk from vector-borne, waterborne, air-borne diseases, pollutants, allergens, and heat effects. Research provides more information about heat events than other climate risks, showing that both physiological differences and summertime play, particularly in vigorous outdoor sports, put teenagers and older children at risk for heat-related consequences. Athletes engaged in sporting events are vulnerable to excess heat and may experience heat-related symptoms such as heat cramps, heat exhaustion, and heat stroke (15). Greater awareness and policies that mandate cooling breaks, water intake, and playing times can be effective protective measures for older children during high heat events.

Children are at particular risk for a variety of adverse mental health effects, especially after experiencing climate disasters. Because children are constantly developing, their reactions will vary by age and maturity as well as the severity of the disaster and the degree to which they experience the disaster, the level of destruction of the built and natural environment, their support systems, the adaptive capacity of their caregivers, and availability and quality of social support systems. Children may experience mental health consequences such as stress, anxiety, depression, PTSD, and suicidality (10, 16). Exposures over the course of childhood can have a cumulative impact on a child's health and well-being (17).

Climate Change Disasters: Exploring the First and Worst for Children

Children's health risks from climate change are greatest for those in marginalized and immigrant communities, indigenous communities, ethnically and racially diverse communities, and for children in poverty, and children of color. Air pollution and heat waves do disproportionate harm to these vulnerable children. In urban areas, heat differentials in neighboring communities are often attributed to the lingering effects of redlining policies as the manifestation of structural racism, which sees these poor and minority neighborhoods experience on-going negative concentration of polluting industries and fewer trees and parks. Consequently, a Black child has nearly twice the risk of developing asthma and almost five times the mortality rate from asthma than a non-Hispanic white child (18).

Climate-related disasters have a disproportionate effect on the physical and mental health of marginalized children. The quality of the built environment where children spend time, especially in their homes, childcare centers, and schools, is critically important for their ability to survive disaster situations. Children whose parents or caregivers have limited funds, limited social networks, and limited options for obtaining energy, transportation, and food during an emergency are more vulnerable as a baseline health condition, and in disaster situations these vulnerabilities are magnified (14). For example, during Hurricane Katrina in 2005, 30% of Black New Orleanians did not have a car to leave the city and therefore suffered disproportionately from the consequences of the resulting floods there (14, 19).

Preparing for disasters is often left to individuals and families to navigate on their own, and access to timely information, emergency funds, and reliable transportation are often not available for everyone. Unfortunately, a general lack of appreciation that children have unique vulnerabilities in the face of disasters is common, therefore, additional policy focus, national and local preparation and recovery planning, and clear public health messaging are critically needed. Recovery from disaster events is also all too often a matter of self-sufficiency, so poorer families have fewer resources and less reserve to prepare for and recover from disasters. Children may suffer physically and emotionally during the disaster as well as in its aftermath, therefore, public health guidance on helping children cope with the physical and mental health impacts of disaster events is most helpful, particularly when messages and interventions are specific to children's developmental stages (20).

Building Equity

As the current trend in climate change continues, environmental health disparities will continue and multiply, with each storm, wildfire, heatwave and drought of increasing intensity and frequency, repeatedly demonstrating the outsized vulnerability of children in poor and marginalized communities (14). As the numbers of disaster events continue to multiply, the risks of society normalizing them, and thereby failing to take protective actions and to provide post-disaster recovery efforts, grows. The increasing pace of climate change places untold stress on fragile communities, endangering children in unprecedented numbers and with consequences that society does not yet fully accept or understand. Is the repetitive nature of climate disasters numbing society? Is normalization of disasters going to become a default and tragic adaptation strategy? Alternatively, the inevitability of continuing climate disasters can inspire political will, corporate largesse, and unprecedented partnerships to launch community level initiatives and actions to contain climate change, build resilience and offset harmful exposures to improve health outcomes for the world's most vulnerable children. Focusing on the most vulnerable will benefit all children, all of humanity and the planet's ecosystems.

The Climate Equity Collaborative

To address the issue of climate change and its differential effects on marginalized children, the Climate Equity Collaborative (CEC) was launched in 2022 (21). The CEC, in collaboration with the Children's Environmental Health Network (CEHN), proposes a program called *Climate Justice for Kid's Health* that expands children's environmental health protections globally and focuses on climate equity. This program plans to connect the world in a technically innovative, collaborative global network, building on CEHN's 30 years of children's health promotion, child-focused public policy, research stimulation, building awareness, advocacy, and education (22). This global network and public-private partnership are devoted to assuring climate justice for children's health, as well as establishing a meeting place to learn and teach, and a platform open to all interested in the intersection of climate change, children's health, and justice. *Climate Justice for Kid's Health* will provide information for the curious, a stage for new findings, with actionable advice for people determined to protect children, ensure health in a changing climate, and pursue livable environments as a human right.

Together, these organizations envision a global movement centered on children in marginalized communities empowered to develop strategies to protect children from climate change. The CEC and CEHN embrace Break the Cycle of Climate Change (BCCC) a program that offers extended one-on-one research mentorship and academic study to minority college and graduate students interested in finding solutions to climate change and health equity dilemmas (5). BCCC bolsters the credentials and understanding of individuals who participate and connects young scholars with a community of dedicated environmental health leaders. The program is modeled after a long-running seminal program, Break the Cycle of Children's Environmental Health Disparities (23).

CEC focuses on youth engagement in climate change equity and is defined by work in two key areas, (i) environmental education, and (ii) children's environmental health. It has the explicit intention to nurture youth leaders and provide youth in marginalized and under-resourced communities with information, ideas for mitigation and adaptation, policy levers, platforms for advocacy, and connections with leaders in a variety of relevant disciplines, all of which to develop ways to join the workforce with environmental health credentials. There are many opportunities to engage the technology that today's children know so well, to nurture learning and critical thinking skills, and providing avenues to technology sector employment for better solutions in energy, housing, and transportation. Minority youth who have experienced climate change disruption, chaos, and pain are well suited to articulate how society's response can protect health, prepare communities for climate change, and find better ways to build resilience and adaptive capabilities that promise tomorrow's children hope for a better future.

At the United Nations Conference of Parties (COP 27) in Egypt in November 2022, the CEC supported grassroots environmental justice activists to speak, network, and influence policy at the Environmental Justice Pavilion (24). Over the last two decades, the evidence, quantification, research, and calls for justice have expanded. But so have the human health harms of climate change. Over these two decades a whole generation of children were born, grew up, and are now young adults. Their disillusionment with global leaders in the halls of power is real and they will continue to let the world know they want change. Young people want to find ways to be included in decision making and to upset the norms that enable perpetuation of a society that continues to operate in a system with potential catastrophic negative consequences (25).

The framework of the CEC is intentionally inclusive, as a public-private partnership which recognizes the need for sectors of society to work together on climate change challenges given the profound implications of environmental exposures and adverse health effects. The CEC originated with the recognition of the limitations of any single public institution or economic sector, and the critical need to build more opportunities for effective community level youth engagement. The CEC can explore the complex elements that naturally collide in a changing climate, by working with federal entities in health, housing, transportation, energy, agriculture, and education through interagency climate change working groups. Engaging with nongovernmental organizations and with the private sector is a well-trod path for addressing big efforts – and climate change undeniably calls for a whole of society approach. CEC will operate through programming to expand meaningful and age-appropriate environmental education and to promote understanding of and protection for children's health. In this way, the CEC can facilitate the disruption of ingrained systems that perpetuate the status quo on environmental threats that increase disparities in environment-related health risks for children.

Youth Expression and Leadership

Most children and youth still have limited voices in global affairs, with little policy influence and no opportunity to vote. And yet, some youths are finding ways to take on the immense problems of their day, to speak up, to act out, and to be heard. Youth-led climate groups are increasing in number and effectiveness. The *Fridays for Future* campaign mobilized over 7 million students in over 6000 climate protests in 185 countries (26). Youths are engaged in stories, poems, art, dance, film, music, and social media about climate change. They engage in science, research, and technology, and they find ever-creative new ways to capture attention and advocate for immediate action. To capture the spirit, energy, and potential of these young people the UN Secretary General has established a Youth Advisory Group on Climate Change (27). Young people are also turning to the courts to plead their cause, making powerful arguments in over 25% of all climate litigation cases filed (28). Other promising actions to promote youth leadership in climate change and health are evident in academic institutions, with programs such as *C-*

Change at Harvard's Chan School of Public Health (29) and *Agents of change in environmental justice* at Columbia University (30).

Adults often fail to realize how acutely children experience trauma, and how they express their fears, especially as these expressions are highly variable by age and developmental stage. A survey of 10,000 young people from age 16 to 25 found that over 45% were worried about climate change such that it affected their daily life and functioning; 59% responded that they were extremely worried, and 84% were at least moderately worried (31). The prevailing perception that the government response to climate change is inadequate, further fuels this anxiety.

Engaging Healthcare Providers

Almost 30 years ago, an Institute of Medicine report recognized that health care providers, particularly obstetricians, pediatricians, and public health professionals, had limited information and training to address environmentally driven health effects, both in the clinical setting and in public health (32). Shortly thereafter, the National Pediatric Environmental Health Unit (PEHSU) network was established and funded by the federal Agency for Toxic Substances and Disease Registry (ATSDR) and the Environmental Protection Agency (EPA), to provide information, consultation, and education around children's environmental health to professionals and to the public. Since then, the network of 10 units based at universities in the 10 US federal regions has amassed a cohort of academics and professionals who have contributed knowledge and understanding of environmental threats to children in consultations, presentations, publications, and in the cultivation of future leaders (33). The PEHSU network continues to provide leadership in training of health professionals in addressing threats to children's environmental health by climate change (34) and environmental injustice (35).

Enhancing awareness and knowledge of clinicians and the public health community on the pediatric impact of climate change and the disparity in effects on children living in poverty and in marginalized communities, is critical to ensure proper care for children. Without such training, critical issues such as management of heat exposure, impacts of natural disasters, or wildfire smoke exposures that affect children with asthma, or allergy control practices for longer and more intense allergy seasons, are likely to be overlooked. *Climate Resources for Health Education (CRHE)* is a useful resource that provides free, publicly accessible, evidence-based resources on climate change and planetary health for health professionals and others seeking to protect children's health in the 21st century (36).

Conclusion

Climate change is an intersectional and multigenerational issue. Focusing on the health of children provides solid and actionable information to ensure that all today's children can safely enter the world, learn, grow, and thrive into adulthood. Infusing environmental equity into the equation is essential – for living with and potentially solving the climate issue is not possible without considering the plight of those most vulnerable and most significantly affected.

The challenge to protect children's health from the effects of harmful environmental exposures is currently a small niche in pediatrics, public health, and environmental protection. However, as we have demonstrated, it is a concern of major proportion and climate change presents a new frontier for children's environmental health, calling for an expanded all-of-society approach centered on youth engagement and leadership to promote justice and protect health.

References

[1] Gamble JL, Balbus J, Berger M, Bouye K, Campbell V, Chief K, et al. Populations of concern. In: Crimmins AJ, Balbus JL, Gamble CB, Beard JE, Bell D, Dodgen RJ, et al., eds. The impacts of climate change on human health in the United States: A scientific assessment. Washington, DC: US Global Change Research Program, 2016:247–86. http://dx.doi.org/10.7930/J0Q81B0T.

[2] Etzel RA, Balk SJ. Children's unique vulnerabilities to environmental hazards. In: Etzel RA, Balk SJ, eds. Pediatric environmental health, 4th ed. Itasca, IL: American Academy of Pediatrics, 2019:17-31.

[3] Sheffield PE, Landrigan PJ. Global climate change and children's health: threats and strategies for prevention. Environ Health Perspect 2011;119:291-8.

[4] Claudio L. Impact of poverty, disparity, and injustice on children's environmental health. In: Landrigan PJ, Etzel RA, eds. Textbook of children's environmental health. Oxford: Oxford University Press. 2014.

[5] Rubin IL, Witherspoon NO. Climate change, environmental justice and children's health: Break the cycle of climate change by cultivating future leaders. J Appl Res Child 2021; 12(1):7.

[6] Philipsborn RP, Chan K. Climate change and global child health. Pediatrics 2018;141(6):e20173774.
[7] UNICEF, The climate crisis is a child rights crisis URL: https://www.unicef.org/media/105376/file/UNICEF-climate-crisis-child-rights-crisis.pdf.
[8] Etzel RA, Balk SJ, eds. Pediatric environmental health, 4th ed. Itasca IL: American Academy Pediatrics, 2019.
[9] Helldén D, Andersson C, Nilsson M, Ebi KL, Friberg P, Alfvén T. Climate change and child health: A scoping review and an expanded conceptual framework. Lancet Planetary Health 2021;5(3):e164-75.
[10] Anderko L, Chalupka S, Du M, Hauptman M, Climate changes reproductive and children's health: a review of risks, exposures, and impacts, Pediatr Res 2020;87:414–9.
[11] Holm SM, Miller, MD, Balmes, JR. Health effects of wildfire smoke in children and public health tools: A narrative review. J Exposure Sci Environ Epidemiol 2021:31:1–20.
[12] Rosa MJ. Identifying critical windows of prenatal particulate matter (PM2.5) exposure and early childhood blood pressure. Environ Res 2020;182:109073.
[13] Auger N, Fraser WD, Smargiassi A, Kosatsky T. Ambient heat and sudden infant death: A case-crossover study spanning 30 years in Montreal, Canada. Environ Health Perspect 2015;123(7):712-6.
[14] Rubin IL, Falk H, Mutic A. Natural disasters and vulnerable populations: A commentary. Int J Child Health Hum Dev 2019;12(4):303-18.
[15] Kerr ZY, Casa DJ, Marshall SW, Comstock RD. Epidemiology of exertional heat illness among US high school athletes. Am J Prev Med 2013;44(1):8-14.
[16] Dodgen D. Mental health and well-being. The impacts of climate change on human health in the United States: A scientific assessment. Washington, DC: US Global Change Research Program, 2016:217-46.
[17] Perera F, Nadeau K. Climate change, fossil-fuel pollution, and children's health, New Engl J Med 2022;386(24):2303-14.
[18] Gutschow B, Gray B, Ragavan MI, Sheffield PE, Philipsborn RP, Jee SH. The intersection of pediatrics, climate change, and structural racism: Ensuring health equity through climate justice. Curr Probl Pediatr Adolesc Care, 2021;52(6):101128.
[19] Bronars S, Ozimek A. Ten years after Katrina: African Americans in New Orleans and the evacuation divide. URL: https://www.forbes.com/sites/modeledbehavior/2015/08/30/ten-years-after-katrina-african-americans-in-new-orleans-and-the-evacuation-divide/?sh = 7138d9b04482.
[20] CDC. Helping children cope with emergencies. URL: www.cdc.gov/childrenindisasters/helping-children-cope.html.
[21] Launch of the climate equity collaborative. URL: https://blog.nwf.org/2022/11/launch-of-the-climate-equity-collaborative/.
[22] Children's Environmental Health Network. URL: https://cehn.org/.
[23] Break the Cycle of Health Disparities Inc. URL: https://www.breakthecycleprogram.org/btcprogram.
[24] United Nations. Climate action. COP 27 Delivering for people and the planet. URL: https://www.un.org/en/climatechange/cop27.

[25] O'Brien K, Selboe E, Hayward BM. Exploring youth activism on climate change: dutiful, disruptive, and dangerous dissent. Ecology Soc 2018;23(3):42.
[26] de Moor J, Uba K, Wahlström M, Wennerhag M, De Vydt M, eds, Protest for a future II: Composition, mobilization and motives of the participants in Fridays for future climate protests on 20–27 September 2019, in 19 Cities around the World, 2020. URL: https://osf.io/3hcxs/download.
[27] United Nations. Climate action. The Youth Advisory Group on climate change. URL: https://www.un.org/en/climatechange/youth-in-action/youth-advisory-group.
[28] Donger E. Children and youth in strategic climate litigation: Advancing rights through legal argument and legal mobilization. Transnational Environ Law 2022;11(2):263-89.
[29] Harvard TH Chan School of Public Health. C-Change at science for a healthy, just, and more sustainable world. URL: https://www.hsph.harvard.edu/c-change/about-c-change/.
[30] Agents for Change in Environmental Justice. URL: https://agentsofchangeinej.org/about/.
[31] Marks E, Hickman C, Pihkala P, Clayton S, Lewandowski E, Mayall EE, et al. Young people's voices on climate anxiety, government betrayal and moral injury: A global phenomenon. URL: https://ssrn.com/abstract=3918955 or http://dx.doi.org/10.2139/ssrn.3918955.
[32] Pope A, Rall P, eds. Committee on Curriculum Development in Environmental Medicine. Environmental medicine: Integrating a missing element into medical education, Washington, DC: National Academies Press, 1995.
[33] Pediatric Environmental Health Specialty Units. URL: https://www.pehsu.net/.
[34] Philipsborn RP, Sheffield P, White A, Osta A, Anderson M, Bernstein A. Climate change and the practice of medicine: Essentials for resident education. Acad Med 2021;96(3):355-67.
[35] Rubin IL, Merrick J, eds. Environmental justice and the intersection of poverty, racism and child health disparities. New York: Nova Science, 2021.
[36] Climate Resources for Health Education (CRHE). URL: https://climatehealthed.org.

Chapter 6

A Survey to Inform Clinical Approaches to Address Climate Health and Equity

Eric Balaban*, MD
and Lesa Walker, MD, MPH

Department of Emergency Medicine, University of Colorado School of Medicine, Aurora, Colorado, United States of America
Environmental Defense Fund, United States Regional Office, Austin, Texas, United States of America

Abstract

Climate change is a present and growing threat to public health. Unfortunately, underrepresented populations experience a disproportionately increased risk to their health. This exploratory survey draws insights from United States-based healthcare providers who identify as part of and/or serve underrepresented communities. The survey assesses their clinical perspectives and practices in addressing the health impacts of climate change (climate-health) and associated health inequities. The survey was designed using input from healthcare leaders with expertise and experience serving underrepresented communities. It was disseminated via partnering clinical provider networks. Results were collated within the SurveyMonkey platform and analyzed using descriptive statistics. The surveyed medical providers recognized climate

* ***Correspondence:*** Eric Balaban, MD, Fellow in Climate Change and Health Policy Science, University of Colorado School of Medicine, Assistant Clinical Professor, Department of Medicine, Faculty at the University of Pittsburgh Medical Center. Email: eric.balaban91@gmail.com; and Lesa R. Walker, MD, MPH, Mentor, Climate Change and Health Policy Science Fellowship, University of Colorado School of Medicine, Medical Consultant, Environmental Defense Fund.
Email: lrwalker04@yahoo.com.

In: Environmental Health: Leaders for Climate Change
Editors: I Leslie Rubin and Joav Merrick
ISBN: 979-8-89530-347-4

change and its potential impacts on patient health but were least familiar with its impacts on health inequities. Only 26% routinely screened for climate change-related health risks, and only 20% discussed these risks in clinical encounters. The most common barriers to screening were time constraints (72%) and competing priorities (53%). Participants suggested that a screening tool could be optimized by prioritizing expediency of use (70%) and incorporation into the electronic medical record system (67%). Conversing with patients about climate change and their health was most hindered by time constraints (68%) and the lack of familiarity with solutions to offer (53%). This study identifies opportunities to optimize educational, material, and systems-level interventions to expand climate-conscious clinical practice. Importantly, these results focus on the lived experiences of frontline providers to inform solutions.

Introduction

Climate change has been described as the single biggest health threat facing humanity (1). As the globe warms, many environmental changes are directly threatening human health. For example, the risk of heat-related illness is increasing for populations on much of the planet (2). The risk of cold weather illnesses, including hypothermia and frostbite, is also increasing as the weather becomes more erratic (3). Intensifying extreme weather will worsen flooding which results in drowning and injury (4). Wildfires (5), air quality (6), and infectious disease (7) are all worsening with climate change as well. In addition to the new illnesses that these disasters will create, chronic diseases are more likely to exacerbate (8). The threat of climate change, along with society's inconsistent response to it, is also a driver of worsening mental illness including anxiety and depression (9).

Other threats to human health may be less direct. For example, a changing climate disrupts predictable crop yields. The nutritional composition of some crops may change as well, and this can elevate the risk of malnourishment among those with tenuous access to food (10).

Food shortages are already forcing human migration, including regions such as the 'Dry Corridor" in Central America. The resulting immigration can strain the stability of neighboring societies charged with supporting the victims of climate disasters. The societal stress created by climate-induced migration can stoke domestic and international tensions that feed societal instability and conflict. Ultimately, climate migration worsens public health by introducing trauma, straining resources, and disrupting continuity of care for those already suffering from disease (11). This potential for climate change to worsen a broad combination of existing health threats has also been acknowledged by the United States Department of Defense, which describes climate change as a "threat multiplier" (12). Importantly, these threats to human health will only worsen as the climate continues to destabilize.

Climate change clearly jeopardizes human health, but it poses a particular risk to underrepresented and disadvantaged populations (13,14). Underrepresented populations in the United States include those that have been historically disenfranchised, under-resourced, or marginalized. This includes underrepresented groups such as Black, Indigenous, and People of Color (BIPoC), Hispanic/Latinx, Asian American, Pacific Islander, immigrants, refugees, and those of low socioeconomic status (SES) (15). Other underrepresented groups may include those who are pregnant, are chronically ill, or are very old or very young. It should be noted that many individuals and communities may carry multiple risk factors, leaving them particularly vulnerable.

Underrepresented groups already tend to experience living conditions that undermine their health. Taken with other adverse social determinants of health, these conditions result in increased acute illness, exacerbations of chronic illness, and limited access to medical care (16). When the increased burden of medical illness is combined with limited access to social or medical support, underserved communities experience diminished climate resilience (17). Undermined climate resilience leads to a reduced ability to anticipate, prepare for, and respond to challenges posed by climate change and its various manifestations.

Pediatric physiology makes children even more susceptible to environmental hazards. Children have higher metabolisms, developing immune systems, and may be unable to seek safety or nutrition depending on their age. Subsequently, children are at a higher risk of illness due to excess heat, unsafe air quality, and nutritional deficiencies (18). A child's susceptibility to climate-driven illness can catalyze a cycle of disparities (see Figure 1). When children become ill, they have an increased need for medical care, which may pose logistical and financial burdens on the family. These challenges may be significant, particularly if the family has limited resources, and may compromise living conditions, job prospects, or financial flexibility, which increases the family's susceptibility to further mental and physical health consequences (19).

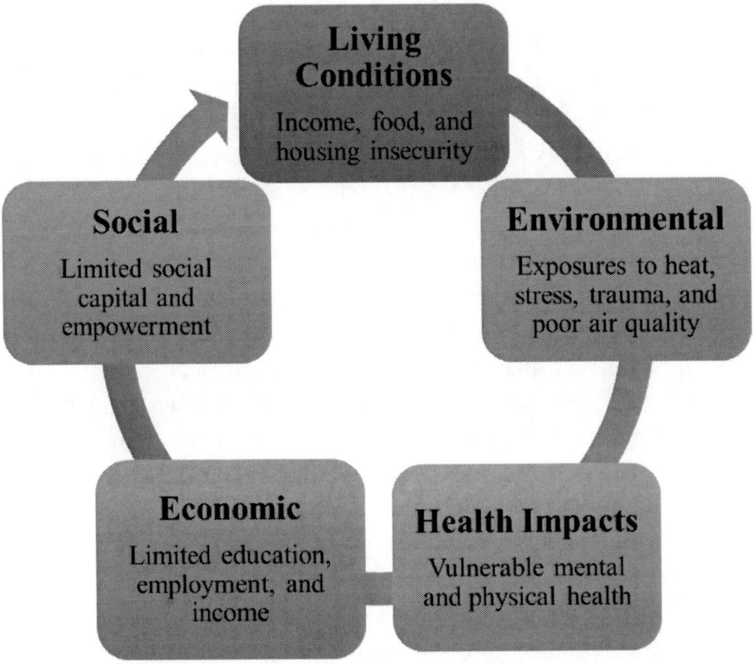

Figure 1. The Cycle of Environmental Health Disparities, which demonstrates the interconnectedness of living conditions, environmental factors, physical health, economic power, and social capital. When healthcare providers effectively address climate-health and equity in clinical practice, they help break this cycle by reducing patient exposure to climate change-related health risks by preventing and treating the health impacts of climate change. Healthcare provider guidance and advocacy can also help implement measures to limit and reduce environmental hazards to health. The yellow stars indicate these focal points in breaking the cycle.

Medical providers are uniquely positioned to disrupt this cycle of disparities by adopting a clinical approach which recognizes and addresses the impacts of climate change on health. Such clinical practice has been described as "climate change-informed care." General guidance has been published for pediatricians (20), but climate change-informed care is applicable to virtually any discipline within the healthcare delivery system. Key elements of such care include implementing environmental health screening, discussing lifestyle modifications, issuing preventative guidance, and providing clinical management of conditions caused by and/or worsened by climate change.

Medical professionals are also uniquely able to advocate for action to address climate change given their firsthand experiences in caring for those whose health is impacted. Providers can offer insight on how changes in the environment increase various risks to patient health. As advocates for their patients, providers are also able to add a valuable, patient-centered perspective to initiatives that call for proactive climate action.

However, there are also several requisite conditions before a healthcare provider can meaningfully address the health impacts of climate change in clinical practice or advocate in the community. Providers must first recognize and acknowledge climate change as a reality and understand its profound impact on human health and health equity. Then, there may be limited availability or application of resources, tools, and training that can hinder the provider's ability to act on that knowledge. Interactive feedback from experienced healthcare providers is essential in adapting clinical practice to address climate-health and equity. Such feedback offers opportunities to build on their successes, identify needs, and overcome barriers.

Since underrepresented communities are disproportionately vulnerable to the ill effects of climate change (13, 14), it follows that healthcare providers who identify with and/or serve those communities are the vanguard of adapting medical practice to meet the growing threat of climate change to human health. These providers have valuable lived experiences that can inform the development of effective clinical practice strategies and standards.

Several prior surveys have attempted to quantify the extent to which medical providers have embraced climate change as a relevant health risk and its related barriers (21-24). Aside from one survey distributed by the National Medical Association (21), none have emphasized the barriers to addressing the health impacts of climate change in clinical practice among underrepresented communities. There are also no known studies that have explored the extent to which providers acknowledge the impacts of climate change on health equity.

This study was conducted as part of the Break the Cycle of Environmental Health Disparities (BTC) program (25). The survey focuses on elevating the voices of medical providers who have lived experiences with underrepresented communities. We surveyed providers who either identify as part of and/or serve underrepresented communities. The survey aimed to measure their familiarities, attitudes, and practices regarding climate change and the impacts of climate change on health and health equity. The survey respondents identify barriers that impede their ability to address climate-health and equity in clinical practice and offer suitable solutions. By gaining their insights, we can optimize care and help "break the cycle" of health disparities related to climate change in patients of all ages.

Our Survey

To conduct an exploratory survey to elevate the voices of a unique subset of healthcare providers, the authors began by establishing working relationships with researchers and healthcare professionals within underrepresented groups throughout the United States. The authors devoted extensive time to identify, contact, communicate with, and strengthen personal connections with 1) experts in health equity, public health, and survey research; and 2) lead administrators and medical directors of professional healthcare organizations and clinical service systems representing and serving underrepresented populations. These connections resulted in a list of project partners, including a broad array of individuals, organizations, and healthcare service providers, to assist with survey development and distribution to their networks (see Acknowledgements). Establishing these partnerships was an essential and highly valued outcome of this project.

During the project design phase, the authors identified and engaged a group of expert advisors to help create the survey. They also gained valuable insights from colleagues who had experience in conducting research or enacting surveys investigating the intersection of climate change, health equity, and clinical practice. The survey was drafted to measure the participant's demographics, their self-assessed familiarity with climate change, its effect on patient health, its effect on health inequities, and the barriers and challenges to screening or adopting clinical practice strategies to address climate-health.

The draft of the survey underwent several rounds of revision based on guidance from the advisors. Their input helped optimize the survey's content, scope, length, and approach. The final survey included 38 questions, four of which were open-ended. A Survey Monkey questionnaire was constructed for survey dissemination.

The survey required participants to acknowledge their consent and complete an assessment of participant inclusion criteria. To emphasize the experiences of medical providers working with underrepresented populations, participants were required to meet both inclusion criteria of 1) being a practicing healthcare provider (physician, nurse practitioner, nurse, physician assistant/associate, dentist) within the United States and 2) identify as part of an underrepresented group and/or serve this population as a significant part of their medical practice. For the purposes of this study, underrepresented was defined as Black, Indigenous, and People of Color (BIPoC); Hispanic or

Latinx; or low SES defined as those earning less than 200% of the poverty level.

The Institutional Review Board (IRB) at the University of Colorado reviewed the survey and granted the project an exempt status. Following IRB approval, the survey was given to the established project partners for electronic dissemination to their healthcare provider networks. These provider networks included, but were not limited to, community health centers, professional medical organizations of healthcare providers identifying with underrepresented groups, medical faculty from Historically Black Medical Schools, Health Equity Physician Fellows, and healthcare clinics/service networks serving underrepresented populations.

The survey was distributed between February 14-25, 2022, a timeframe required to meet overarching BTC project deadlines. Results were collected within a private SurveyMonkey account and analyzed using descriptive statistics with Microsoft Excel software. In addition to analyzing the general results of all completed surveys, a subgroup assessment specific to partner network demographics was conducted and shared with that network.

Our Findings

177 surveys were collected over a 12-day period. 27 surveys were excluded due to self-identified inability to satisfy the inclusion criteria. An additional 32 surveys were excluded due to being incomplete, yielding 118 serviceable surveys.

Notable demographic findings of survey respondents include a majority of female respondents (64%, $n = 76$) and a fairly even distribution of age (see Table 1). 45% ($n = 53$) identified as white, non-Hispanic/Latinx healthcare practitioners. Black-identifying participants constituted 22% ($n = 26$), and Hispanic/Latinx-identifying participants constituted 15% ($n = 18$) of survey respondents.

Participants were predominantly physicians (80%, $n = 94$) from 27 different states. A significant majority of responses were collected from providers in Texas and California (65%). Providers were primarily outpatient-based (72%, $n = 85$). Urban, suburban, and rural areas were represented nearly equally. Family medicine practitioners and pediatricians together represented a majority (71%, $n = 84$) of 14 different types of practices that responded to the survey.

Table 1. Demographics of survey respondents

Demographic	Number of Respondents	Percentage of Respondents
Overall	118	100%
Gender		
Female	76	64%
Male	40	34%
Non-binary	1	1%
Deferred	1	1%
Race		
Asian or Asian American	15	13%
Black or African American	26	22%
Indigenous Peoples or Native American	5	4%
White	71	60%
Other	1	1%
Ethnicity		
Hispanic or Latinx	18	15%
Non-Hispanic or Latinx	100	85%
Age		
18-34	16	14%
35-44	34	29%
45-54	25	21%
55-65	31	26%
>65	12	10%
Profession		
Physician	94	80%
Advanced Practice Provider	21	18%
Nurse	3	3%
Home State*		
Texas	44	37%
California	33	28%
Maryland	6	5%
Iowa	4	3%
Kentucky	3	3%
Nevada	3	3%
Illinois	2	2%
Nebraska	2	2%
North Dakota	2	2%
Tennessee	2	2%
Practice Setting		
Urban	48	41%
Suburban	28	24%
Rural	42	36%
Practice Base		
Outpatient	85	72%
Inpatient	1	1%
Outpatient and Inpatient	32	27%

Demographic	Number of Respondents	Percentage of Respondents
Practice Specialty^		
Family Medicine	66	56%
Pediatric Medicine	18	15%
Internal Medicine	18	15%
Obstetrics and Gynecology	3	3%
Dentistry	3	3%
Surgery	2	2%

*States with one respondent not listed, including Alaska, Arizona, Colorado, Georgia, Idaho, Kansas, Minnesota, Montana, New Jersey, New Mexico, Ohio, Oklahoma, Oregon, Utah, Virginia, Washington, Wyoming.

^Specialties with one respondent not listed, including Dermatology, Emergency Medicine, Psychiatry, Counseling, Addiction Medicine, Oncology, Occupational Medicine.

Participants reported a significant presence of underrepresented patients constituting their clinical practice. Sixty-four percent (n = *76*) estimated that greater than half of their practice serviced such groups (see Figure 2). Patients of low SES had the greatest representation in participants' clinical practice (86%, n = *102*), with strong representation of Hispanic/Latinx and BIPoC groups (74% and 62%, respectively) as well (see Figure 3).

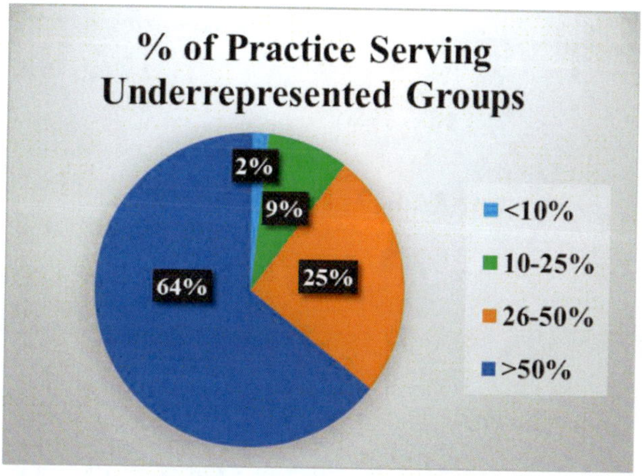

Figure 2. Breakdown of approximate practice time devoted to serving underrepresented groups of patients, where 'underrepresented' is defined as Hispanic/Latinx, Black, Indigenous, People of Color, and people of low socioeconomic status for the purposes of this study.

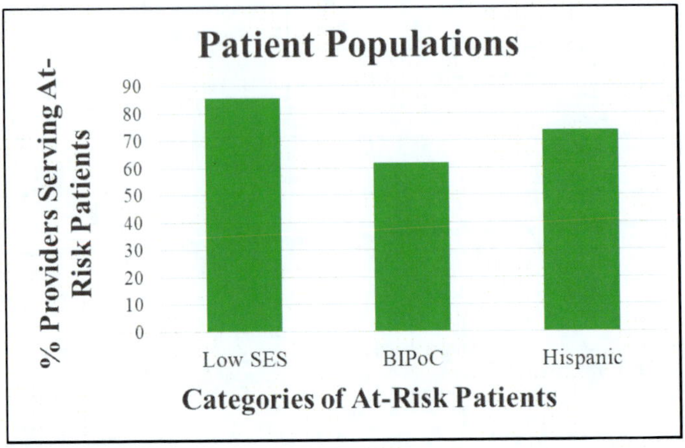

Figure 3. Breakdown of providers who report serving an underrepresented patient population as a significant part of their clinical practice. (Low SES = low socioeconomic status; BIPoC = Black, Indigenous, People of Color).

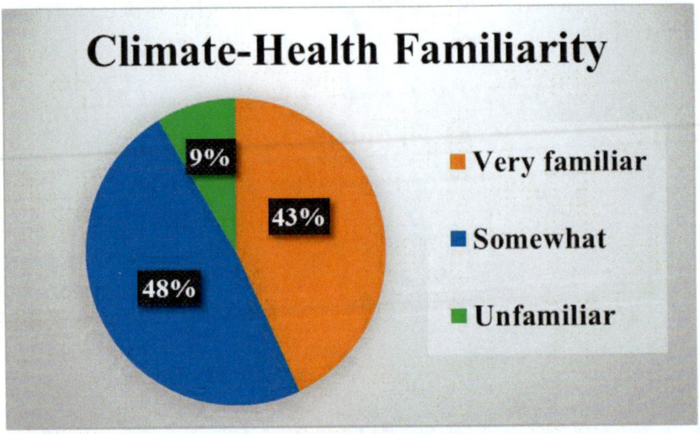

Figure 4. Self-assessed (by the respondent) degree of familiarity with the adverse health risks caused by climate change.

A striking number (99%, *n* = *117*) of participants felt either 'Very' or 'Somewhat' familiar with the phenomenon of climate change. A slightly smaller but still majority of participants felt 'Very' or 'Somewhat' familiar with the possibility that climate change can impact patient health (91%, *n* = *108*) (see Figure 4). Twenty-six percent (*n* = *31*) noted climate change-related health concerns in half or more of their clinical encounters. The most frequent concerns were related to poor air quality (67%, *n* = *79*), housing security (55%, *n* = *65*), and psychiatric illness (58%, *n* = *49*) (see Figure 5).

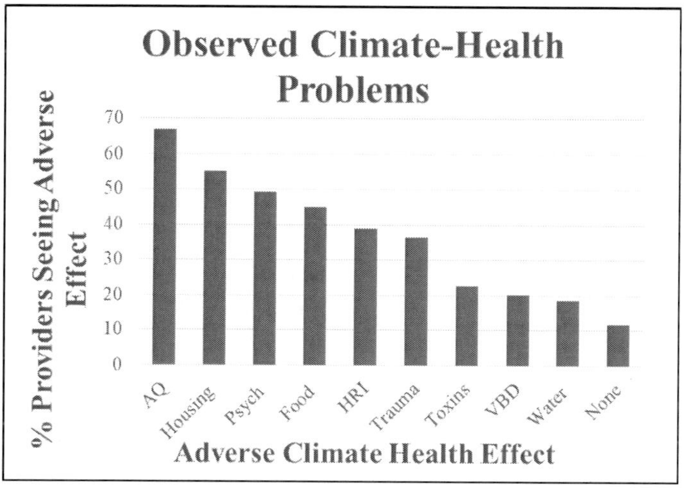

Figure 5. Negative health impacts related to climate change reported by medical Providers. (AQ = air quality-related illness; Housing = housing insecurity; Psych = mental illness or behavioral issues; Food = food insecurity; HRI = heat-related illness; Toxins = toxin exposure; VBD = vector-borne disease; Water = water insecurity).

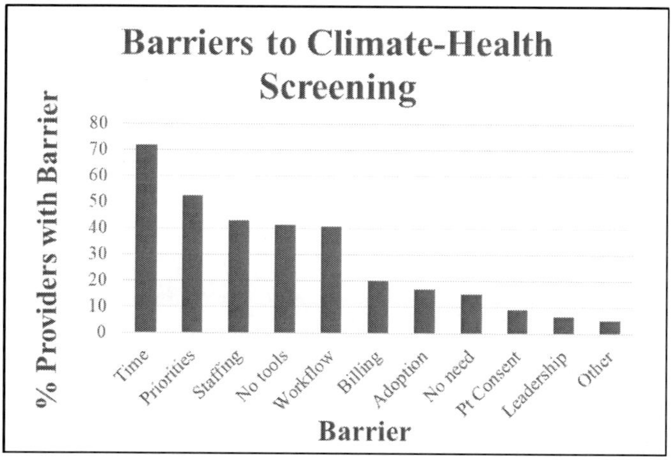

Figure 6. Frequency of barriers encountered by medical providers in enacting routine screening for health risks related to climate change. (Priorities = clinical priorities; Staffing = lack of staff or financial support; No tools = lack of available screening tools; Billing = inability to bill for climate-health counseling; Adoption = providers not adopting screening practices; Pt Consent = patient not consenting to counselling; Leadership = lack of leadership approval).

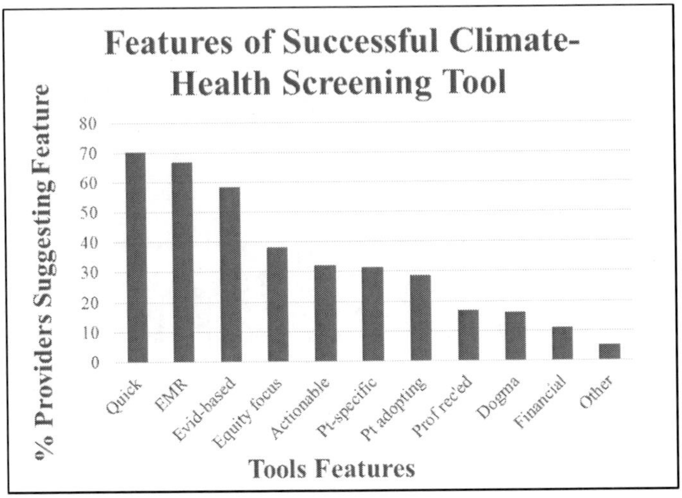

Figure 7. Features reported by medical providers that would make a tool used to screen for climate change-related health risks most successful. (EMR = incorporated into the electronic medical record system; Evid-based = evidence-based to improve public health; Equity focus = intentioned to serve those socially, racially, or ethnically underrepresented; Pt-specific = specifically relevant to the provider's patient population; Pt adopting = perceived importance by the patient; Prof rec'ed = recognized or recommended by a professional organization; Financial = connected with financial incentive for the clinician).

When asked if routine environmental health screening took place as part of their clinical practices, only 26% ($n = 31$) said 'Yes'. Screening was left to respondent interpretation according to their own screening standards. Reported screening modalities included screening for only a few specific domains of climate-health as part of routine medical history gathering. The most common examples include screening for food insecurity, depression, and lead exposure.

Time was the most common barrier to implementing screening (72%, $n = 85$). Other significant barriers to screening included climate change's impact on health not being an institutional priority (53% $n = 62$), the lack of an existing screening tool (42%, $n = 49$), and the lack of support staff to implement screening tools (43%, $n = 51$) (see Figure 6). Commonly suggested important features of a successful climate-health screening tool included its ability to be used quickly (*68%*, $n = 80$), its ability to interface with the electronic medical record system (67%, $n = 79$), and the tool being evidence-based or validated (58%, $n = 69$) (see Figure 7).

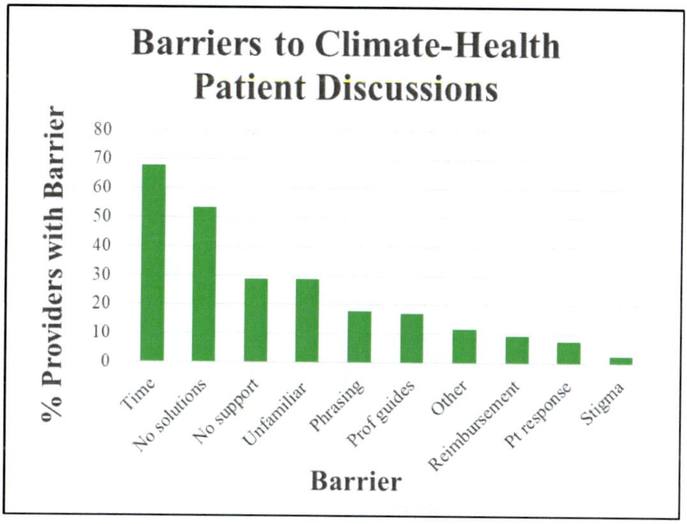

Figure 8. Frequency of barriers encountered by medical providers discussing the health risks related to climate change with patients. (No solutions = Lack of familiarity with solutions to offer; No support = lack of supporting resources; Unfamiliar = lack of familiarity with the topic; Phrasing = unsure how to broach or phrase; Prof guides = lack of professional society guidance; Reimbursement = lack of service reimbursement; Pt response = fear of poor patient response; Stigma = fear of stigma from colleagues).

Providers infrequently discussed the topic of climate change and its impact on health with patients, with 80% (*n* = *94*) reporting they rarely or never broach the topic. The most common barriers cited included time constraints (68%, *n* = *80*) and not being familiar with actionable interventions or solutions (53%, *n* = *63*) (see Figure 8). Additionally, only 27% (*n* = *32*) reported being familiar with how to obtain more information on climate change's effect on health.

According to survey respondents, patients were even less likely to ask about climate change and its relevance to their health. 93% (*n* = *109*) of respondents reported patients rarely or never ask. However, patient reactions to discussions on climate change and health were generally reported as positive or neutral (94%) (see Figure 9).

Participants were least familiar with climate change's impacts on health inequities, with 75% (*n* = *83*) reporting being 'Somewhat Familiar' or 'Unfamiliar' with the topic (see Figure 10). It was also the least discussed in clinical encounters, with 85% (*n* = *94*) reporting they rarely or never engage in the topic with patients. When asked the importance of the provider's race in conducting such conversations, responses were mixed. However, nearly all participants (99%, *n* = *109*) felt it was important to use a communication style that acknowledged and respected a patient's culture.

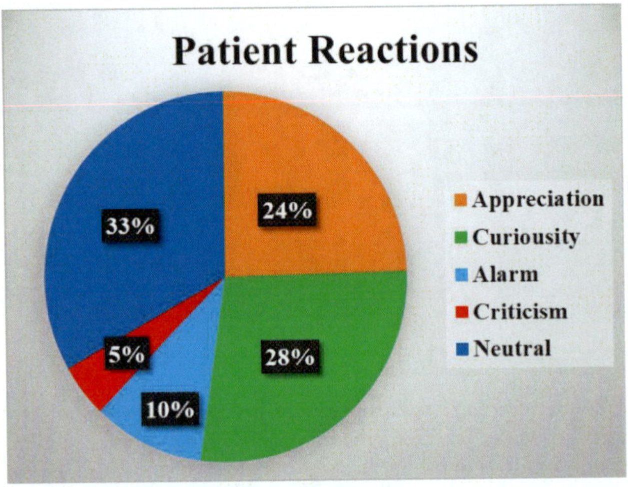

Figure 9. Reactions patients exhibited during discussions of climate change and health as reported by healthcare providers. (Criticism = patients responding negatively or skeptically).

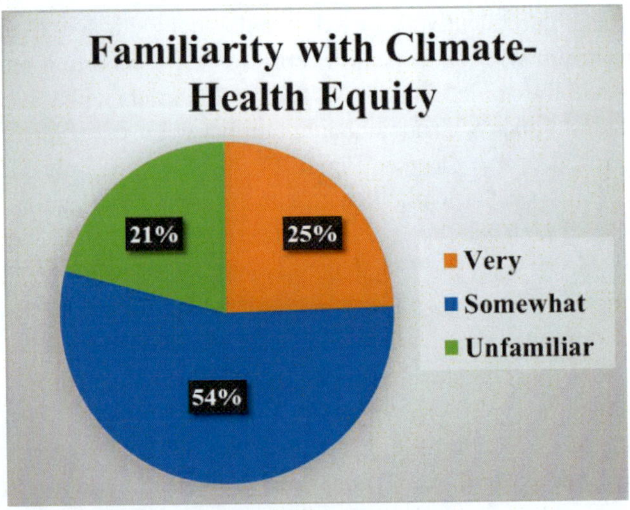

Figure 10. Self-assessed degree of familiarity with how the adverse health Risks caused by climate change contribute to health inequity.

Discussion

This exploratory survey successfully focused on perspectives from healthcare providers who identify as being part of and/or serve underrepresented communities as indicated by several demographic markers. First, the composite of respondents' race and ethnicity captured a high representation of minority groups. Additionally, most providers reported that BIPoC patients, Hispanic/Latinx patients, and patients of low SES had a significant presence in their practice with over half of respondents estimating that 50% or more of their practice was constituted by such populations.

The survey found that medical providers among underrepresented communities were familiar with climate change and identified climate change as a relevant factor for their patients' health. This finding is similar to the results of other large surveys conducted among United States physicians (21-24).

Additional findings show that climate change-related health concerns are already prominent among these communities. Roughly three out of four providers reported connecting poor patient health and climate change in at least a few clinical encounters. When identified, poor air quality, mental health or behavioral issues, and insecurities in housing or food availability were the most prevalent climate change-related health concerns. Over half of survey participants noted each of these concerns as being common, active climate change-associated health risks they observed.

Although climate change-related health risks were a well understood and prominent concern, relatively few providers reported screening for these risks. Approximately three out of four providers did not routinely screen for climate change-related health risks among their patients. The most common modality of climate change-related health risk screening implemented by providers was part of routine medical history gathering. This relatively unstructured modality of screening can include elements of environmental risk factors, such as depression or food insecurity, though is significantly less comprehensive than a screening process dedicated specifically to environmental- or climate change-related health risk factors.

Several significant barriers to screening were identified. Time constraints proved to be the most significant factor in limiting providers from routine screening. This emphasizes how busy these providers already are. Providers serving underrepresented communities may be particularly busy given the complexity of social challenges faced by their patients. Other system-level challenges included lack of staffing and available clinical screening tools.

These findings help pinpoint some of the systemic difficulties encountered by clinicians in addressing climate-health and equity in their clinical practice.

This survey illuminated several key aspects that would optimize a tool to facilitate climate change screening with patients. Respondents felt that a tool should be quick to use, validated using-evidence-based medicine, and should interface with the electronic medical record system. These findings are particularly valuable, as they are suggestions offered by future users of such a tool.

Relatively few providers broached climate change within the context of medical discussions with patients. Over three out of four providers discussed climate change and health in 'a few encounters' or never, with time constraints again proving to be a powerful barrier. Providers also reported a significant need for more education on the health risks that climate change poses and practical solutions to offer.

Importantly, healthcare providers who discussed climate change reported that most patients reacted with appreciation or curiosity while very few reacted critically. This study also found that very few patients initiated such conversations with their health care providers, emphasizing the important role of healthcare providers in addressing this topic.

Finally, this survey found that medical providers' familiarity with climate change's effect on health inequities was not as robust as their familiarity with climate change and its impact on health alone. This finding is particularly important given that the survey selected for providers who are most likely to interface with communities experiencing health inequities compounded by climate change. This feedback suggests the need for additional clinical education on the health disparities associated with climate change, with emphasis on the clinician's role in addressing these disparities and advocating for environmental justice.

Strengths and Limitations

A strength of this study is the focus on developing partnerships with healthcare professionals and networks identifying as and/or serving underrepresented communities. The survey highlights the perspectives of these healthcare providers and recognizes the importance of their lived experience to inform clinical practice strategies in addressing the impacts of climate change on health. By sharing the survey results with the project's partners, these groups directly benefited from the findings. The survey is the beginning of a valuable conversation. It calls attention to the requisite and fundamental importance of engaging underrepresented communities to develop clinical practice strategies that address climate-health and equity.

Although the survey results appear to align with some conclusions drawn from prior works, the results are subject to biases inherent to optional survey-based research. This survey was designed to select for its target population while being freely disseminated among multiple networks, so it is also impossible to estimate an accurate response rate. This study's inclusion of several underrepresented groups, combined with a limited sample size, lends itself to considerable heterogeneity in respondents as well.

The generalizability of these findings may be limited by the fact that the large majority of respondents originated from Texas and California. This skew in respondent geography is likely a result of several partner organizations being based in those states. The skew may be somewhat mitigated by the remaining one-third of respondents originating from numerous other states. However, it must be noted that the reported survey results in over-represented responses from providers in the two predominant states.

It should also be noted that some responses rely on self-reporting of respondent familiarities with complex topics, which may or may not reflect the respondent's objective depth of understanding. Additionally, as exploratory research, the survey results are descriptive and do not offer statistical significance, however, they do provide valuable insight and suggest targets for more in-depth studies and subsequent interventions.

Subsequent works may expand on these results by involving other underrepresented populations not included here. Future studies that are larger or offer more narrative detail through focus group interviewing would expand the scope of the findings. As intended, there is extensive opportunity for future studies to build on these results to optimize clinical practice interventions to address climate-health and equity. Finally, future works may adopt the approach used in this project to form partnerships with underrepresented groups of interest. It is only through this crucial, ongoing effort that the benefits of such collaborations on the delivery of healthcare and health equity related to climate change will be realized.

Conclusion

The clinical encounter with a patient provides a critical window of opportunity for healthcare providers to "break the cycle" of health disparities related to climate change. They can screen for health risks, offer anticipatory guidance, prescribe safety measures, identify and treat health impacts of climate change, and provide ongoing patient education and resources. Overcoming barriers in clinical practice to address the health impacts and disparities due to climate change should be considered a priority for healthcare providers. In identifying and overcoming such barriers, it is crucial to seek out and listen to the voices of those who are most impacted by climate change.

This exploratory survey focused on input from healthcare providers who identify with and/or serve underrepresented groups. These providers reveal barriers to addressing climate-health and equity in their clinical practices. Their responses suggest targeted interventions to improve screening tools, patient and provider education, and clinical practice strategies. Though the survey results are limited and not generalizable to all healthcare providers, the results do represent the reality of those who responded. Thus, there are existing frontline providers who need additional support and who also have ideas for quality improvements in clinical practice. Their input offers valuable insights to transform clinical practice so that it is maximally effective in addressing the threats of climate change and in breaking the cycle of environmental health disparities.

Acknowledgments

Foremost, we acknowledge and thank our partners who collaborated in the construction and editing of this survey, including Dr Roxana Cruz of the Texas Association of Community Health Centers, Dr Leonore Okwara at the Association of Black Researchers, and Dr Tonyka McKinney and Dr Shaneeta Johnson at the Satcher Health Leadership Institute as part of the Morehouse School of Medicine. We thank them for their insight during the study design process and for volunteering their networks as avenues of survey distribution.

We also thank Dr Leslie Rubin, Founder of the Break the Cycle program and faculty at Morehouse School of medicine, Dr Mark Mitchell within the National Medical Association, Dr Lori Byron within the Indian Health Service and American Academy of Pediatrics, Ms Angelina Renteria within the Indian Health Council, and Dr Gary Greensweig and Dr Ankita Sagar within CommonSpirit Health for distributing the survey among their networks as well.

We acknowledge Dr Linda Rudolph and Dr Aaron Bernstein for offering their input regarding climate change-related surveys and Ms. Nicole Hill with ecoAmerica for offering her insight regarding survey design and deployment. Their contributions helped optimize the survey design.

We also thank the Break the Cycle of Health Disparities program for supporting and guiding the efforts of this project, as well as the opportunity to present these findings at the 2022 Break the Cycle Conference.

We are grateful to the University of Colorado School Of Medicine's Fellowship in Climate Change and Health Policy for providing mentorship in addition to funding for the SurveyMonkey platform, and we thank the Environmental Defense Fund for providing consultation support to the Fellowship and this project.

References

[1] World Health Organization. Climate change and health. (URL: https://www.who.int/news-room/fact-sheets/detail/climate-change-and-health.

[2] Duffy PB, Tebaldi C. Increasing prevalence of extreme summer temperatures in the US. Climatic Change 2012;(2)111:487-95.

[3] Vose RS. Monitoring and understanding changes in extremes: Extratropical storms, winds, and waves. Bull Am Meteorological Soc 2014;3(95):377-86.

[4] Doocy S, Daniels A, Murray S, Kirsch TD. The human impact of floods: A historical review of events 1980-2009 and systematic literature review. PLoS Curr 2013;5: ecurrents.dis.f4deb457904936b07c09daa98ee8171a.

[5] Reidmiller DR, Avery CW, Easterling DR, Kunkel KE, Lewis KLM, Maycock TK, Stewart BC. Impacts, risks and adaptation in the United States: Fourth National Climate Assessment 2018;2. doi: 10.7930/NCA4.2018.

[6] Centers for Disease Control and Prevention. Climate change decreases the quality of the air we breathe. URL: chrome-extension://efaidnbmnnnibpcajpcglclefindmkaj/https://www.cdc.gov/climateandhealth/pubs/air-quality-final_508.pdf.

[7] Rocklov J, Dubrow R. Climate change: an enduring challenge for vector-borne disease prevention and control. Nat Immunol 2020;21:479-83.

[8] Centers for Disease Control and Prevention. Climate and health. URL: https://www.cdc.gov/climateandhealth/effects/default.htm.

[9] Cianconi P, Betro S, Janir L. The impact of climate change on mental health: a systematic descriptive review. Front Psychiatry 2020 Mar 06. https://doi.org/10.3389/fpsyt.2020.00074.

[10] Swinburn BA. The global syndemic of obesity, undernutrition, and climate change: The Lancet commission report. Lancet 2019:393(10173):791-846.

[11] The White House. Report on the impact of climate change on migration. Washington, DC: The White House, October 2021.

[12] Banusiewicz JD. Hagel to address threat multiplier of climate change. Washington, DC: US Department of Defense, 2014. URL: https://www.defense.gov/News/News-Stories/Article/Article/603440/.

[13] Department of Health and Human Services. Climate change and health equity, and environmental justice at HHS. Washington, DC: Department of Health and Human Services, Office of Climate Change and Health Equity, 2022. URL: https://www.hhs.gov/climate-change-health-equity-environmental-justice/index.html.

[14] Environmental Protection Agency. EPA report shows disproportionate impacts of climate change on socially vulnerable populations in the United States. URL: https://www.epa.gov/newsreleases/epa-report-shows-disproportionate-impacts-climate-change-socially-vulnerable.

[15] Rudolph L, Harrison C, Buckley L, North S. Climate change, health, and equity: A guide for local health departments. Oakland, CA and Washington, DC: Public Health Institute and American Public Health Association, 2018. URL: https://www.apha.org/-/media/files/pdf/topics/climate/climate_health_equity.ashx.

[16] Satcher D. Eliminating racial and ethnic disparities in health: The role of the ten leading health indicators. J Natl Med Assoc 2000;92:315–8.

[17] Center for Climate and Energy Solutions. Climate essentials. What is climate resilience and why does it matter? URL: chrome-extension://efaidnbmnnnibpcajpcglclefindmkaj/https://www.c2es.org/wp-content/uploads/2019/04/what-is-climate-resilience.pdf.

[18] American Academy of Pediatrics Council on Environmental Health. Global climate change and children's health. Pediatrics 2015;136(5):992-7.

[19] Rubin IL, Geller RJ, Martinuzzi K, Howett M, Gitterman BA, Wells L, et al. Break the cycle of environmental health disparities: An ecological framework. Int Public Health J 2017;9(2):115-30.

[20] Philipsborn RP, Cowenhoven J, Bole A, Balk SJ, Berstein A. A pediatrician's guide to climate change-informed primary care. Curr Probl Pediatr Adolesc Health Care 2021; 51(6):101027.

[21] Sarfaty M, Mitchell M, Bloodhart B, and Maibach EW. A survey of African American physicians on the health effects of climate change. Int J Environ Res Public Health 2014; 11(12):12473-85.

[22] Sarfaty M, Bloodhart B, Ewart G, Thurston GD, Balmes JR, Guidotti TL, Maibach EW. American Thoracic Society member survey on climate change and health. Ann Am Thorac Soc 2015;12(2):274-8.

[23] Sarfaty M, Kreslake JM, Casale TB, Maibach EW. Views of AAAAI members on climate change and health. J Allergy Clin Immunol Pract 2016;4(2):333-5.

[24] Kotcher J, Maibach E, Miller J, Campbell E, Alqodmani L, Maiero M, et al. Views of health professionals on climate change and health: a multinational survey study. Lancet 2021;5(5):316-23.

[25] Break The Cycle of Health Disparities Inc. URL: https://www.breakthecycleprogram.org/.

Chapter 7

Historical Residential Redlining and Children's Vulnerability to Climate Change in the Southeast United States

Devon Nenon[1],
Yui Fujii[2], BS, MPH
James Martin III[1], MS, BS
Christine C. Ekenga[3], MPH, PhD
and Eric S. Coker[1,*], BS, MS, PhD

[1]Department of Environmental and Global Health, College of Public Health and Health Professions, University of Florida, Gainesville, Florida, United States of America
[2]Department of Biostatistics, College of Public Health and Health Professions, University of Florida, Gainesville, Florida, United States of America
[3]Gangarosa Department of Environmental Health, Rollins School of Public Health, Emory University, Atlanta, Georgia, United States of America

Abstract

Little is known about the impacts of historical redlining, a racially discriminatory housing practice, on future or current cumulative children's environmental health burdens. The Southeast has one of the hottest climates in the United States (US) and the severity of extreme heat is expected to increase in the region due to climate change. The southeast also faces high levels of concentrated child poverty and air pollution-related inequalities. This study investigates relationships between

* ***Correspondence:*** Eric S. Coker, 1225 Center Dr, Room 4159, Gainesville, FL 32610, United States.
Email: eric.coker@phhp.ufl.edu.

In: Environmental Health: Leaders for Climate Change
Editors: I Leslie Rubin and Joav Merrick
ISBN: 979-8-89530-347-4

redlining and combined stressors for multiple dimensions of the environment, including differences in climate change-related temperatures, in Alabama, Florida, Georgia and South Carolina. Methods: We combined data on long-term historical ambient $PM_{2.5}$ and maximum temperature, climate change-related temperature projections, greenspace (normalized differential vegetative index [NDVI)) and child poverty. Multilevel linear regression and cluster analysis tested associations between Homeowners' Loan Corporation grades (A through D) and multiple environmental stressors, including differences in child poverty and climate change-related temperatures. Results: Relative to grade A neighborhoods, redlined neighborhoods are associated with higher historical maximum temperature ($\beta = 0.07$ [95% CI: 0.03, 0.11]), average temperatures in 2080 ($\beta = 0.03$ [0.00, 0.05]), summertime temperatures in 2080 ($\beta = 0.06$ [0.03, 0.09]), and changes in summertime temperatures ($\beta = 0.006$ [0.002, 0.01]). Average $PM_{2.5}$ ($\beta = 0.24$ [0.16, 0.32]) and NDVI ($\beta = -0.06$ [-0.08, -0.04]) are significantly worse in redlined neighborhoods. Cluster analysis identified redlined neighborhoods vulnerable to heat-related climate change impacts and combined high child poverty and environmental stressors. Conclusions: Historical redlining is associated with climate-related heat burdens in the Southeast US. This study can inform targeted climate adaptation efforts for communities impacted by structural racism.

Introduction

Climate change has important implications for human health (1). The environmental hazards associated with climate change include increased temperatures, reduced air quality, and more frequent and severe weather extremes. Direct exposure to these hazards has the potential to contribute to the development and exacerbation of adverse health conditions, while the indirect effects of climate-related environmental hazards involve disruptions to critical infrastructure systems, limited access to healthy food and clean water, and poor access to high-quality health and social services (2). The extent of both the direct and indirect health impacts of climate change is likely influenced by the exposome, a complex mix of environmental stressors, including natural, built, and social stressors (3, 4). A comprehensive understanding of the exposome is critical to advancing our knowledge regarding the health impacts of climate change.

Epidemiologic studies suggest that children's health is adversely affected by extreme heat temperatures and air pollution. From the prenatal period to early childhood, acute and chronic exposure to elevated temperature events and ambient $PM_{2.5}$ exposure have all been observed to have associations with adverse birth outcomes, emergency department visits, and chronic health outcomes (5–8). Although all children are vulnerable to environmental stressors, some children are more vulnerable than others due to disproportionately elevated exposures and other factors, such as poor access to health care services, underlying chronic diseases (e.g., asthma), poor nutrition, or psychosocial stressors (e.g., racial discrimination). In the United States (US), for example, there are well-documented race- and place-based disparities in exposure to harmful natural, built, and social stressors, and the environmental justice movement emerged in response to the disproportionate exposure of poor, minoritized, and disenfranchised populations to environmental chemical and non-chemical stressors (9). In the context of climate change and children's health, these exposure disparities may play a role in differential vulnerability to the extreme heat temperatures and air pollution exposures associated with climate change, particularly for poor and minority children in the US.

Those in the Environmental Justice Movement have long advocated for considering multiple environmental non-chemical and chemical stressors in the context of broader structural factors, which comprise economic, social, policy, and organizational elements, that manifest in environmental injustices (10). In the US, historical residential redlining, a discriminatory housing practice, is a prime example of a structural factor that has resulted in decades of economic inequalities and segregation in urban areas. Redlining was practiced by the Home-Owner's Loan Corporation (HOLC), a federal agency created in 1933, as part of Roosevelt's New Deal to stabilize the housing market during the Great Depression. The HOLC offered loans to bail out struggling financial institutions and refinanced mortgages for homeowners, but these loans were not equally available to all. They created color coded maps that assigned letter grades of A, B, C, or D to neighborhoods based on racial demographics, and lower graded areas were frequently denied home loans and economic investment, a practice that continued through the 1960s. This was rationalized as a metric for "mortgage security" but instead was based on explicitly racist metrics, mirrored in the language used to justify the grades, such as "colored infiltration." The maps were color-coded, with A ("best") grades colored green, B ("still desirable") colored blue, C ("definitely declining") colored yellow, and D ("hazardous") colored red; hence the term "redlining" (11).

Redlining, therefore, is a form of historical and ongoing structural racism, one that is embedded into the organizations and patterns of our society, and it has been linked to environmental stressors, such as extreme heat temperatures, and adverse child health outcomes, such as low birth weight, (12). Moreover, as others have shown, there is a high correlation and plausible causal relationships in the US between low socioeconomic status and residential racial segregation with several of the housing, multipollutant exposures, and neighborhood amenities mentioned above (13–17). For instance, redlining played a key role throughout many American cities in creating unequal access to housing and immobility for Black communities. One recent study showed that children residing in redlined neighborhoods experienced substantial joint social inequalities that include lower educational attainment, reduced income in adulthood, higher poverty rates, and higher rates of single-headed households (18). Therefore, there is a need to consider the implications of historical redlining on current and future climate change-related environmental stressors. Several US studies have found that historically redlined neighborhoods often have fewer neighborhood-level amenities like green spaces to help mitigate against extreme high temperature events (19) However, to our knowledge, no study has considered all three dimensions of the exposome - natural, built, social - in a climate-focused analysis. The simultaneous evaluation of natural, built, and social environmental stressors provide a more comprehensive and holistic assessment of the influence of historic redlining on children's vulnerability to climate change (20).

The purpose of this study was to examine the relationship between historical residential redlining, current-day exposure to natural, built, and social environmental stressors, and future heat-related climate impacts in four Southeastern (SE) states in the US (Alabama, Georgia, Florida, and South Carolina (see Figure 1A). Our study focused on these four SE states because their long history of structural racism, which encompassed enslavement and other forms of marginalization of people of African descent (Figure 1B), as well as region-specific vulnerabilities to climate change (Figure 1C). Recently published data indicated that several SE states, including GA, FL, and SC, have experienced an increase in the absolute number of extreme heat events ("heat waves") compared to 1948 (21). Furthermore, summer-time temperature projections for this region suggest a further increase due to climate change (see Figure 1B), and while several studies suggest climate change will increase ambient $PM_{2.5}$ levels on average across the US (22–24), some predictions indicate that the SE US will see a larger increase in ambient $PM_{2.5}$ compared to other US regions (24). We hypothesize that historical residential redlining would be associated with: 1) current-day exposure to multiple (natural, built, and social) environmental stressors and 2) future exposure to ambient heat temperatures. Further, since it is not enough to merely characterize redlining-based environmental disparities (25), an additional purpose of this study was to provide data that can inform strategies for addressing the environmental and climate-related impacts of structural racism. We approached this by employing a data-driven cluster analysis that identifies the specific redlined communities facing higher levels of childhood poverty, multiple environmental inequalities, racial residential segregation, and projected heat-related impacts. Our approach thus characterizes *and* identifies redlined communities that should be prioritized for addressing environmental injustices and informing local climate change adaptation strategies.

Study

The study area comprises cities in the southeastern (SE) US states of Alabama, Florida, Georgia, and South Carolina (see Figure 1). The 13 SE cities included in the analysis (see Figure 1A) are those with digitized HOLC maps made available in the Mapping Inequality dataset (11). The SE region was selected because of its history of entrenched structural racism and previous research showing environmental vulnerabilities driven by climate change (see Figure 1C), such as increasing summer-time temperatures (see Figure 1A) and PM2.5 air pollution in this region (24). In addition, these four states in the SE were selected based on the availability of PM2.5 land-use regression modeling data (described below).

Data

Selection of study variables is informed by emerging children's environmental health literature on extreme weather events, air quality, built and social environments, and redlining (2-6, 15-19, 21). Social environment datasets include: 1) spatial polygon shapefiles of graded HOLC neighborhoods from the Mapping Inequality geospatial library; 2) a Census tract-level racial segregation index obtained from the Othering and Belonging Institute at the University of California, Berkeley, which measures the divergence of racial demographics within a tract from the demographics at the country or city level and 3) the percentage of children below poverty at the tract-level from the US Census Bureau. Tract-level variables for segregation and percent child poverty were merged with a 2010 US Census tract boundary shapefile in R software environment for statistical computing and graphics using Geographical Information Systems (GIS). These tabular data on census tract-level variables of segregation and child poverty were first transformed into raster data. Average segregation and child poverty raster grid cells were then averaged over HOLC polygons in R to assign an average level of racial segregation and child poverty for each HOLC neighborhood. These HOLC neighborhood-specific values were used in subsequent analyses.

Eight variables were selected to represent different dimensions of the natural and built environments. These natural and built environment factors include the following gridded surface (raster) data:

1. Modeled daily maximum ambient temperature averaged between 2012 and 2016 available in the gridMET data product (4 km by 4 km resolution)
2. Modeled daily maximum relative humidity averaged between 2012 and 2016 available in the gridMET data product (4 km by 4 km resolution)
3. Modeled long-term ambient PM2.5 concentrations averaged between 2012 and 2016 using a monthly land use regression model (described below) (1 km by 1 km resolution)
4. Projected annual average temperature (2080s, RCP 8.5) developed by the US Department of Agriculture (USDA) using MACAv2METDA (4 km by 4 km resolution)
5. Projected absolute change in annual average temperature (historical to future) developed by the USDA using MACAv2METDA (4 km by 4 km resolution)
6. Projected Summer average temperature (2080s, RCP 8.5) developed by the US Department of Agriculture (USDA) using MACAv2METDA (4 km by 4 km resolution)
7. Projected absolute change in summer average temperature (historical to future) developed by the USDA using MACAv2METDA (4 km by 4 km resolution)
8. Normalized differential vegetation index (NDVI) available from the Moderate Resolution Imaging Spectroradiometer (MODIS) (1 km by 1 km resolution).

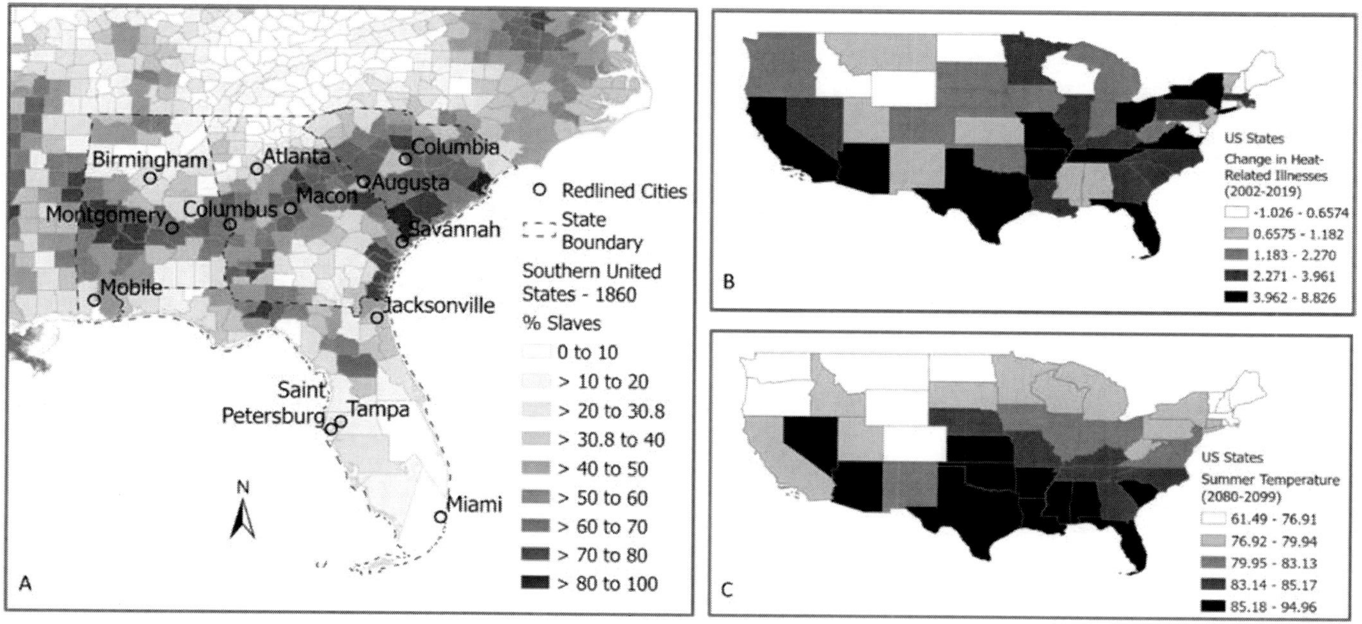

Figure 1. (A) Map showing the Southeastern United States study area including the 13 cities with available HOLC redlining GIS maps. This map portrays the level III ecoregion boundaries established by the US EPA. (B) Map showing the number of slaves within each US state or territory as reported in the 1830 US Census (26). (C) Map showing the projected mean summer-time temperatures (degrees Celsius) in 2080 for States in the conterminous US (27).

These gridded raster values for each environmental indicator were mapped in GIS and averaged over HOLC neighborhood polygons using R as done with the social environment variables.

Data Analysis

We computed summary statistics for all study variables overall and stratified by HOLC grade. We estimated associations between HOLC grade and environmental variables using hierarchal linear regression modeling. A random effect for city was fit in all models. Hierarchical regression models relating HOLC grade with climate variables were adjusted for ecoregion (level III). An additional model (Model 2) included a fixed effect of segregation using a segregation index score.

We estimated $PM_{2.5}$ for each HOLC graded neighborhood in our study area with a LUR model using machine learning (ML) methods adapted from a previously published study (28). Briefly, ML algorithms were trained using a combination of remote sensing (aerosol optical depth and NDVI) and modeled gridded surface data (e.g., weather parameters) and other land-use feature data as independent predictor variables, with ground-level ambient $PM_{2.5}$ air monitoring data as the response variable. We performed a 10-fold cross-validation on random splits with 80% training and 20% test sets for each modeling algorithm (28, 29). The random split was based on observation months for each air monitor. The fitted algorithms include ordinary least squares *(OLS)* regression, Random Forest *(RF),* polynomial support vector machines *(svmPoly),* and k-Nearest Neighbor regression *(KNN).* In addition to these base-learner algorithms, we fit an ensemble model, a linear combination of the four base-learner algorithms (OLS, GAM, RF, KNN). Machine learning (ML) algorithms were fitted using the *caret* package in R. We used the *'caretList'* command to fit all ML algorithms in parallel. The *caretEnsemble* package in R was used to fit the ensemble model. From the cross-validation, we computed root mean squared error (RMSE) and R^2 (coefficient of determination) of the observed versus fitted values to assess the model performance of the various ML algorithms. The best performing model was the ML ensemble model with an R^2 of 0.80 and RMSE of 1.12 $\mu g/m^3$. Trained models were then used to predict ambient $PM_{2.5}$ at unmonitored locations on a 1 km x 1 km gridded surface.

Bayesian Profile Regression

An overarching objective of this study is to identify profiles of within-city differences in socio-environmental burdens and projected ambient temperature from climate change and to explore how these profiles relate to current childhood poverty in the study area. To address this, we applied Bayesian Profile Regression (BPR), a semi-parametric cluster analysis technique. BPR uses an infinite mixture model to identify clusters based on the joint distribution of multiple covariates. Set in a Bayesian framework, the infinite mixture model uses a Dirichlet process with stick-breaking priors, allowing the number of clusters and cluster assignment to vary across Markov Chain Monte Carlo (MCMC) iterations. BPR also fits a sub-model to simultaneously estimate an association (risk) between exposure profile clusters and a response variable. With post-processing of the MCMC output, BPR computes a dissimilarity matrix and uses a partition around medoids approach to optimize cluster assignment. The Bayesian model averaging then averages cluster allocations and risk values available from the MCMC output.

For the cluster analysis, we analyzed city-specific Z-scores at the HOLC grade level that we computed for each natural, built, and social environmental exposures and climate change temperature projections. A Z-score is a commonly used metric in social epidemiology and social sciences to indicate within-region inequalities for specific area-level indicator variables. This standardization is beneficial in the context of our study as we are interested in within-city spatial inequalities and because we are interested in pooling data between regions for cluster analysis (described below). Since each study variable exhibits different city-specific baseline levels for social (e.g., segregation and child poverty), built, and natural environmental factors (e.g., air pollution, temperature, NDVI, etc.), standardization with city-specific Z-scores enables us to perform cluster analysis that identifies neighborhoods with similar and dissimilar spatial inequalities. Precisely, we fit social exposure variables including segregation index score; Natural environment exposures including ambient PM2.5 (2012-2016) and historical maximum temperature (2012-2016). Built environment exposure included NDVI (greenspace). Climate Change temperature projections were used including projected temperature in 2080, the temperature difference in 2080 from historical values, summer temperature in 2080, and summer temperature difference in 2080. All neighborhood-level Z-scores were reclassified into quartiles for the cluster analysis, which we modeled jointly as discrete covariates. The dependent variable in the BPR sub-model was the percentage of children below poverty. Therefore, BPR analysis in our study models the linear association between socio-environmental inequality clusters and the risk of childhood poverty. Clusters were mapped in ArcGIS (ESRI, Redlands, CA) to highlight the spatial patterning of these inequality clusters and the risk of childhood poverty for HOLC grades for cities in our study area.

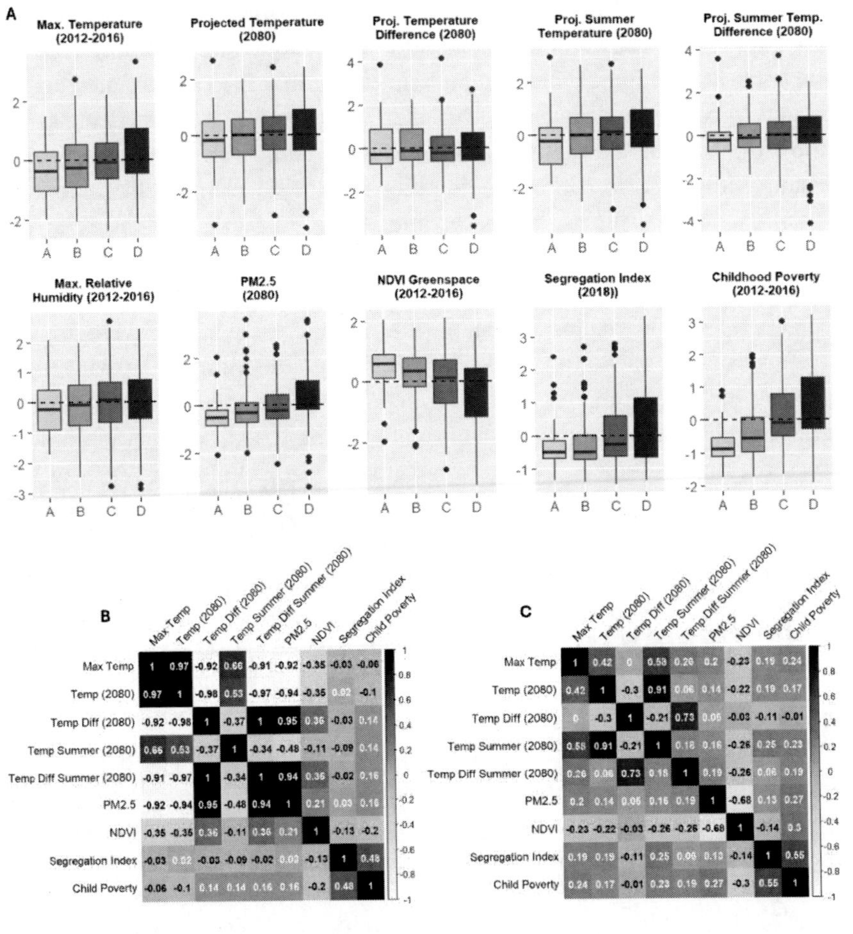

Figure 2. (A) Boxplots of city-specific z-scores for study variables stratified by HOLC grade. (B) Between-variable Pearson's correlations as calculated with average HOLC neighborhood values. (C) Between-variable Pearson's correlations as calculated with city-specific HOLC neighborhood Z-scores.

Findings

The boxplots in Figure 2a display the distributions, stratified by HOLC grade, of the unadjusted social, built, and natural environment Z-scores. We can see worsening Z-scores moving from 'A,' 'B,' 'C,' and 'D' HOLC grades for each environmental factor. A statistical summary of raw values for each environmental variable overall and stratified by HOLC grade is provided in the Supplemental Materials (see Tables S1 and S2). Figures 2b and 2c further summarize the between-variable correlations for the raw and normalized Z-score values. These correlation plots portray the interrelated nature of the study area's social, built, and natural environments. The correlation between several study variables differs when comparing correlations estimated from raw values versus city-specific Z-scores. For example, the study-wide correlations between $PM_{2.5}$ and maximum historical temperature are negative with raw values, whereas Z-scores show a positive correlation between $PM_{2.5}$ and maximum historical temperature. Similarly, raw values of $PM_{2.5}$ and NDVI (surrogate for greenspace) exhibit a positive correlation overall, yet Z-scores of $PM_{2.5}$ and NDVI reveal a negative (opposite) correlation. Notably, both $PM_{2.5}$ and NDVI Z-scores result in positive and negative correlations for all climate-related (historical and projected) variables.

Multilevel linear modeling results are summarized in Figure 3. A detailed summary of unadjusted and adjusted associations is further provided in Table S3 and Table S4 in the supplemental materials. For climate-related response variables, worsening HOLC grade is monotonically associated with significantly ($p < 0.05$) higher historical maximum temperature, higher projected temperature for 2080, higher average summer temperature for 2080, and higher summer temperature difference for 2080. HOLC grade is also associated with significantly higher $PM_{2.5}$, significantly higher childhood poverty, and significantly lower NDVI (greenspace). Independent of HOLC grade, a higher level of segregation is associated with higher maximum historical temperature, higher temperature in 2080, higher summer temperature in 2080, higher summer temperature difference in 2080, higher childhood poverty, and lower NDVI. Increasing childhood poverty level is also associated with significantly higher maximum historical temperature, higher summer temperature difference in 2080, and lower NDVI.

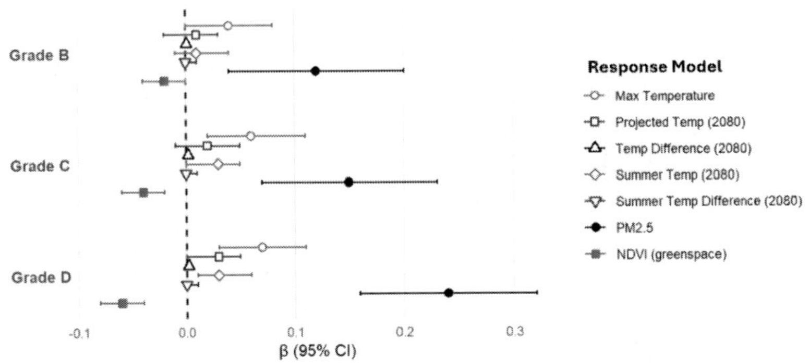

Figure 3. Adjusted associations between environmental factors and HOLC grades. Each regression coefficient corresponds to a separate model specific to each different environmental variable.

Characterization of Exposure Profile Clusters

Based on our study's social, built, and natural environmental factors, Bayesian profile regression detected 10 clusters of exposure profiles. The heat map in Figure 4 shows the distinct exposure profile clusters in terms of the environmental factor-specific Z-score means for each cluster. Clusters with the highest levels of adverse burden of social, built, and natural environmental factors are cluster 8 (C8), cluster 9 (C9) and cluster 10 (C10). These three clusters are the only exposure profiles with positive mean Z-scores for segregation (indicating high within city segregation) and concurrent higher inequality burdens for most, if not all, other built and natural environmental factors. Based on data from the USDA climate change model, these three clusters (C8, C9, and C10) are projected to experience more significant temperature impacts tied to climate change. For instance, neighborhoods in C8, C9, and C10 are projected to experience an increase in their summer-time temperatures in 2080 compared to historical summer-time temperatures. Clusters that appear to be particularly vulnerable to summer-time temperature increases are clusters 7 (C7), C8, and C9.

The same three exposure profiles with adverse exposure profiles—C8, C9, and C10—are each characterized by significantly elevated childhood poverty compared to the baseline (mean) childhood poverty levels of the study population (see Figure 5). The posterior probabilities of childhood poverty above baseline for C8, C9, and C10 are all >0.95. None of the other clusters resulted in a posterior probability of childhood poverty above baseline >0.95. When examining the hard clusters, the empirical median poverty levels for C8, C9, and C10 are 46.2% (interquartile range [IQR] = 30.9, 56.6), 49.0% (IQR = 39.9, 60.8), and 48.6% (IQR = 38.8, 55.9), respectively; well above the overall child poverty levels in the study population (median = 23.6%; IQR = 10.18, 45.4).

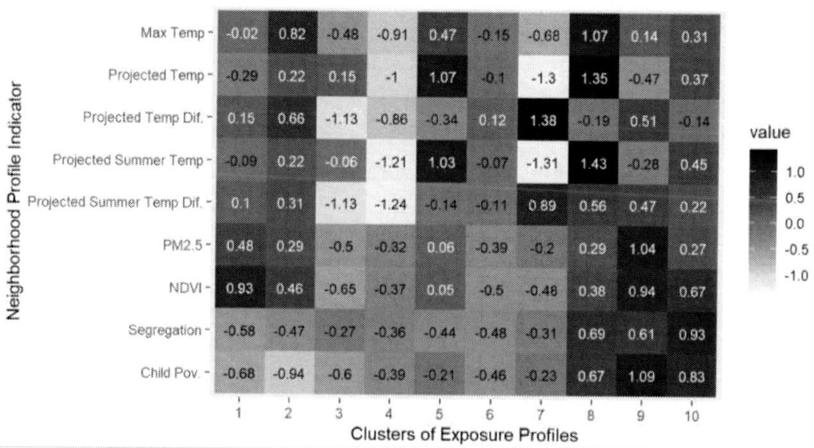

Figure 4. City-specific built, natural, and social-environmental variable Z-scores for each cluster.

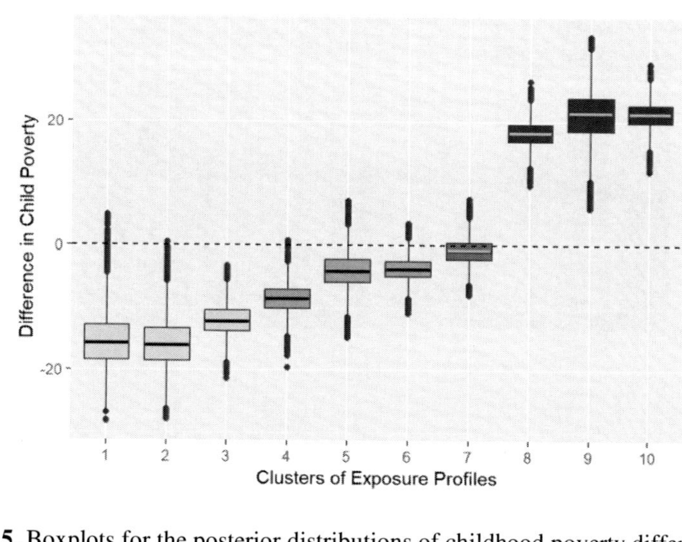

Figure 5. Boxplots for the posterior distributions of childhood poverty differences compared to the study-wide mean childhood percent poverty.

Distribution of HOLC Grades Within Exposure Profile Clusters

The distribution of the percentage of neighborhood HOLC grades within each cluster is summarized as stacked bar graphs in Figure 6. While 61.5% of neighborhoods in the study area are classified as either C or D grades, 75% or more of the neighborhoods within the highest childhood poverty clusters (C8, C9, and C10) are classified as grade C or grade D neighborhoods. Nearly all (94.7%) of the neighborhoods in C9 are either grade D or grade C neighborhoods. Grade D comprises 42.8% (C8), 71.1% (C9), and 37.8% (C10) of the neighborhoods within the three highest poverty levels, whereas grade D comprises only 28% of the neighborhoods in our entire study area. To help illustrate the spatial coincidence of redlined communities and vulnerable clusters, we highlight two major cities in the Southeastern US—Atlanta, GA and Miami, FL (see Figure 7). A complete presentation of maps depicting the spatial pattern of exposure profile clusters within each city included in the analysis is provided in the Supplemental Materials (Figure S1). This cluster analysis empirically confirms that worse HOLC grades face disproportionately higher combined social, built, and environmental stressors and projected increases in temperature due to climate change compared to better HOLC grades.

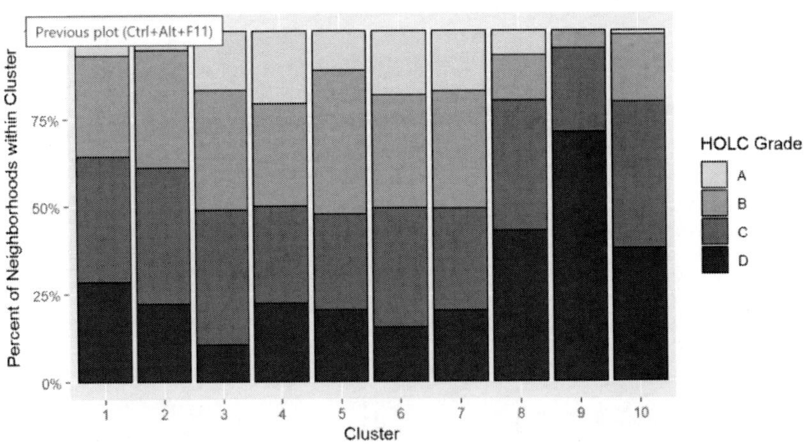

Figure 6. Stacked bar plots portraying the distribution of HOLC neighborhood grades within each cluster.

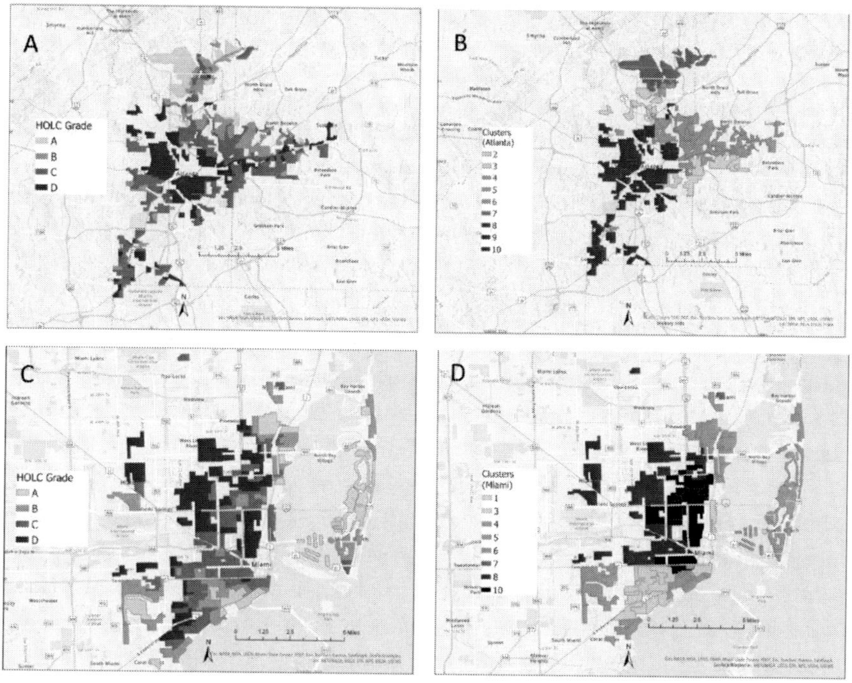

Figure 7. Maps of Atlanta, GA (panels A & B) and Miami, FL (panels C & D) depicting the spatial coincidence of vulnerable clusters (C8, C9, and C10) and worse HOLC neighborhood grades (grades C and D).

Discussion

The purpose of this study is to model the multivariate relationship between historical residential redlining, joint childhood vulnerability to current-day exposure to natural, built, and social environmental stressors, and potential future heat-related climate impacts in four Southeastern (SE) states in the US (Alabama, Georgia, Florida, and South Carolina, see figure 1A). Our main finding is that neighborhoods with higher childhood poverty levels coincide with worse HOLC graded neighborhoods (redlined neighborhoods) facing disproportionately higher combined social, built, and environmental stressors compared to better HOLC grades. Importantly, our analysis reveals that redlined neighborhoods may experience heat-related climate impacts more than neighborhoods that were not redlined.

Ours is the first study, to our knowledge, linking historical redlining maps with combined indicators of social, built, and natural environmental stressors. There are several previous studies (30) that explore the individual relationships between redlining and increased environmental stressors and corroborate our findings. One study found that redlined neighborhoods are particularly vulnerable to the urban heat island effect by comparing land surface temperatures in 108 US cities by HOLC grade (31). They also found that redlined neighborhoods typically have elevated temperatures when compared to their city-specific baseline. Their results were also more pronounced in the Southeast, further supporting the importance of our study area of interest. Another recent nationwide study found that intra-urban variation of ambient NO_2 and $PM_{2.5}$ was explained by redlining patterns in the US, with worse HOLC grades associated with significantly higher pollutant concentrations (30). Current-day greenspace levels across the US was recently shown in another analysis (19) to be significantly lower in worse HOLC grade neighborhoods than in better HOLC grade neighborhoods. Another recent study reported that segregation levels are higher in redlined neighborhoods compared with non-redlined neighborhoods, further corroborating our study findings. We further note that higher childhood poverty levels are shown in our study to be associated with higher levels of intra-urban segregation, $PM_{2.5}$, temperatures, and lower levels of greenspace. Our study, combined with others in the existing literature, demonstrate the important role of redlining in shaping disparities in total environmental stressors in the Southeastern states of Alabama, Georgia, Florida, and South Carolina.

This is also the first study to explore the potential for longer-term climate change disparities tied to residential redlining. While it is not surprising that neighborhoods with higher present-day intra-urban temperatures are projected to have higher temperatures in the future due to climate change, our models suggest that redlined neighborhoods are estimated to have significantly higher differences in Summer-time temperatures in 2080 compared with Summer-time temperature differences for non-redlined neighborhoods. Hence, there is a potential for even more pronounced risk of heat-related events and adverse health outcomes for children in redlined communities compared to their non-redlined community counterparts. Moreover, if present-day joint social and built environment inequalities assessed in our study persist, adaptation and mitigation to such heat-related events induced by climate change could be impaired for many redlined communities in the Southeastern US. We recommend additional research into the potential for redlining-based climate disparities, including other types of climate hazards (e.g., sea-level rise or natural disasters), to better characterize the extent of this problem.

This study also expands on previous work addressing children's environmental health disparities by connecting redlining to environmental risk factors. The environmental and socioeconomic factors explored in this study are known to be connected to environmental health inequalities, and children are particularly vulnerable to these inequalities. Extreme heat is connected to a range of health problems in children, a relationship that is well established in the literature. Previous research has found that after days of extreme heat, there are significant increases in visits to children's emergency departments not only for heat related illness, but also for ear infections, parasitic and infection diseases, nervous system diseases, neurodevelopmental disorders, and other illnesses (32). Our study results are also consistent with previous research connecting both heat, prenatal exposures, and redlining to preterm birth, a health impact that sets children at a disadvantage from the very start of their lives (33, 34). Our results expand on the connection between redlining and these environmental factors to further characterize the relationship between redlining and children's environmental health disparities.

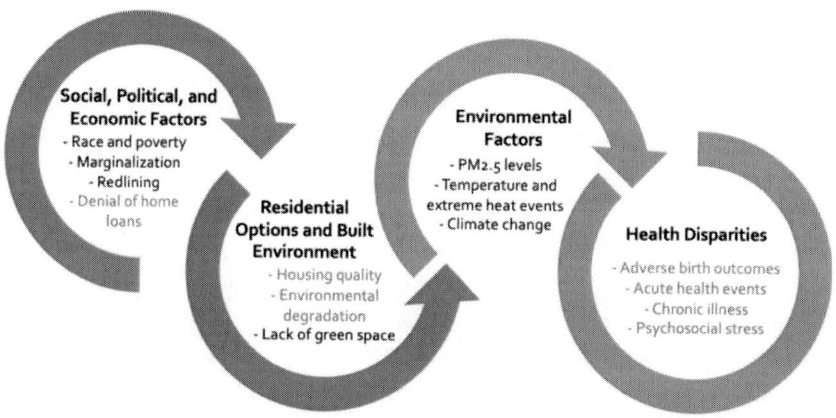

Figure 8. The cycle of environmental health disparities, highlighting the aspects explored in this study

Overall, this study breaks the cycle of environmental health disparities (see Figure 8) by providing an important metric to inform climate adaptation efforts. For instance, this study identifies redlined neighborhoods that face disproportionate burden to specific environmental and social stressors that can be prioritized and targeted for climate adaptation. Further quantifying and establishing the impact of historical redlining on current environmental burdens also gives greater insight into the structural roots of environmental health disparities. Acknowledging these roots and the role of structural racism in current and future climate impacts is vital in breaking the cycle of disparities that impact children in redlined neighborhoods throughout their lives.

Strengths and Limitations

There are key methodological innovations of the current study that relate to the leveraging of a multivariate statistical method designed to assess combined environmental stressors and the mapping of communities facing these joint environmental stressors. Environmental justice communities, which are impacted by historical residential redlining, are theorized to face numerous environmental stressors that can combine in complex ways (30, 35, 36). The application of conventional regression modeling, however, faces several practical barriers in such contexts. For instance, important environmental stressors can be highly collinear, exhibit non-linear relationships or interactions, and such relationships can vary geographically (spatial non-stationarity). With Bayesian profile regression (BPR), we were able to apply a semi-parametric cluster analysis technique that leverages the collinearity of the environmental stressors to identify distinct exposure profiles. This approach also allowed us to map the complex intra-urban spatial patterns of the exposure profiles, which helps avoid issues of non-stationarity when estimating global effects such as the relationship between multiple environmental stressors and childhood poverty over a large geographic area. From a policy and public health action perspective, the mapping of exposure vulnerability profiles, with high combined levels of environmental stressors, can help direct limited resources towards these communities in a more just way. By highlighting the adverse community-level factors that propagate the cycle of children's environmental health disparities, and mapping their locations, our approach goes beyond simply identifying socio-environmental inequalities.

There are some important limitations of our study. While it is critical to highlight the structural role of historical redlining and its ongoing impact to current environmental health disparities and associated future climate risk, there are other place-based aspects of the built, natural, and social environment that we did not investigate. These may include, but are not limited to, additional hazardous air pollutants, housing factors such as access to air conditioning, ventilation, and the age of housing stock, availability of neighborhood amenities like public parks, and other socioeconomic factors like educational attainment and unemployment. Analysis of these other factors may reveal even more complex relationships and inequalities similar to what was revealed in our study. Despite this limitation, the place-based factors analyzed in this study were selected due to the extensive literature showing that these factors are strong determinants of childhood health disparities along with high causal evidence for adverse child health outcomes.

Acknowledgments

The authors thank James Martin III for his data management of spatial data in this study, and the Break the Cycle of Environmental Health Disparities program for their support and assistance.

References

[1] Wilson B. Urban heat management and the legacy of redlining. J Am Plann Assoc 2020;86(4):443-57.

[2] McMichael AJ, Lindgren E. Climate change: Present and future risks to health, and necessary responses: Review: Climate change and health. J Intern Med 2011;270(5):401-13. A

[3] Wild CP. The exposome: From concept to utility. Int J Epidemiol 2012;41(1):24-32.

[4] Juarez P, Matthews-Juarez P, Hood D, Im W, Levine R, Kilbourne B, et al. The public health exposome: A population-based, exposure science approach to health disparities research. Int J Environ Res Public Health 2014;11(12):12866-95.

[5] Ilango SD, Weaver M, Sheridan P, Schwarz L, Clemesha RES, Bruckner T, et al. Extreme heat episodes and risk of preterm birth in California, 2005–2013. Environ Int 2020;137:105541.

[6] Kuehn L, McCormick S. Heat exposure and maternal health in the face of climate change. Int J Environ Res Public Health 2017;14(8):853.

[7] Basagaña X, Michael Y, Lensky IM, Rubin L, Grotto I, Vadislavsky E, et al. Low and high ambient temperatures during pregnancy and birth weight among 624,940 singleton term births in Israel (2010–2014): An investigation of potential windows of susceptibility. Environ Health Perspect 2021;129(10):107001.

[8] Sun S, Spangler KR, Weinberger KR, Yanosky JD, Braun JM, Wellenius GA. Ambient temperature and markers of fetal growth: A retrospective observational study of 29 million US singleton births. Environ Health Perspect 2019;127(6):067005.

[9] Bullard, Robert D. Environmental justice: Its more than waste facility siting. Soc Sci Q 1996;77(3):493-9.

[10] Marshall BD, Kerr T, Shoveller JA, Montaner JS, Wood E. Structural factors associated with an increased risk of HIV and sexually transmitted infection transmission among street-involved youth. BMC Public Health 2009;9(1):7.

[11] Nelson RK, Winling L, Marciano R, Connolly N. Mapping inequality. American Panorama. URL: https://dsl.richmond.edu/panorama/redlining/#loc=5/39.1/-94.58 &text = intro.

[12] Hollenbach SJ, Thornburg LL, Glantz JC, Hill E. Associations between historically redlined districts and racial disparities in current obstetric outcomes. JAMA Netw Open 2021;4(9):e2126707.

[13] Adamkiewicz G, Zota AR, Fabian MP, Chahine T, Julien R, Spengler JD, et al. Moving environmental justice indoors: Understanding structural influences on residential exposure patterns in low-income communities. Am J Public Health 2011;101(S1):S238-45.

[14] Rosofsky A, Levy JI, Breen MS, Zanobetti A, Fabian MP. The impact of air exchange rate on ambient air pollution exposure and inequalities across all residential parcels in Massachusetts. J Expo Sci Environ Epidemiol 2019;29(4): 520-30.

[15] Rauh VA, Landrigan PJ, Claudio L. Housing and health: Intersection of poverty and environmental exposures. Ann NY Acad Sci 2008;1136(1):276-88.

[16] Wen M, Zhang X, Harris CD, Holt JB, Croft JB. Spatial disparities in the distribution of parks and green spaces in the USA. Ann Behav Med 2013;45(S1):18-27.

[17] Rossen LM, Pollack KM. Making the connection between zoning and health disparities. Environ Justice 2012;5(3):119-27.

[18] Aaronson D, Mazumder B, Hartley DA, Harrison Stinson M. The long-run effects of the 1930s redlining maps on children, Rochester, NY, 2022. URL: https://papers. ssrn.com/abstract = 4094372.

[19] Nardone A, Rudolph KE, Morello-Frosch R, Casey JA. Redlines and Greenspace: The relationship between historical redlining and 2010 greenspace across the United States. Environ Health Perspect 2021;129(1):017006.

[20] Senier L, Brown P, Shostak S, Hanna B. The socio-exposome: Advancing exposure science and environmental justice in a postgenomic era. Environ Sociol 2017;3(2):107-21.

[21] Climate change indicators: High and low temperatures. US Environmental Protection Agency, 2022. URL: https://www.epa.gov/climate-indicators/climate-change-indicators-high-and-low-temperatures.

[22] Park S, Allen RJ, Lim CH. A likely increase in fine particulate matter and premature mortality under future climate change. Air Qual Atmosphere Health 2020;13(2):143-51.

[23] Shen L, Mickley LJ, Murray LT. Influence of 2000–2050 climate change on particulate matter in the United States: Results from a new statistical model. Atmospheric Chem Phys 2017;17(6):4355-67.

[24] Nolte CG, Spero TL, Bowden JH, Mallard MS, Dolwick PD. The potential effects of climate change on air quality across the conterminous US at 2030 under three

Representative Concentration Pathways. Atmospheric Chem Phys 2018;18 (20):15471-89.

[25] Levy JI. Invited perspective: Moving from characterizing to addressing racial/ethnic disparities in air pollution exposure. Environ Health Perspect 2021;129(12):121302.

[26] Manson, Steven, Schroeder, Jonathan, Van Riper, David, Kugler, Tracy, Ruggles, Steven. National Historical Geographic Information System: Version 17.0. Minneapolis, MN: IPUMS, 2022. URL: https://www.nhgis.org/.

[27] Projected annual average temperature (2080s, RCP 8.5). US Department of Agriculture. URL: https://data.fs.usda.gov/geodata/rastergateway/OSC/climate.php.

[28] Coker ES, Amegah AK, Mwebaze E, Ssematimba J, Bainomugisha E. A land use regression model using machine learning and locally developed low cost particulate matter sensors in Uganda. Environ Res 2021;199:111352.

[29] Weissert L, Alberti K, Miles E, Miskell G, Feenstra B, Henshaw GS, et al. Low-cost sensor networks and land-use regression: Interpolating nitrogen dioxide concentration at high temporal and spatial resolution in Southern California. Atmos Environ 2020;223:117287.

[30] Lane HM, Morello-Frosch R, Marshall JD, Apte JS. Historical redlining is associated with present-day air pollution disparities in US cities. Environ Sci Technol Lett 2022;9(4):345-50.

[31] Hoffman JS, Shandas V, Pendleton N. The effects of historical housing policies on resident exposure to intra-urban heat: A study of 108 US urban areas. Climate 2020;8(1):12.

[32] Bernstein AS, Sun S, Weinberger KR, Spangler KR, Sheffield PE, Wellenius GA. Warm season and emergency department visits to US childrens hospitals. Environ Health Perspect 2022;130(1):017001.

[33] Cushing L, Morello-Frosch R, Hubbard A. Extreme heat and its association with social disparities in the risk of spontaneous preterm birth. Paediatr Perinat Epidemiol 2022;36(1):13-22.

[34] Krieger N, Van Wye G, Huynh M, Waterman PD, Maduro G, Li W, et al. Structural racism, historical redlining, and risk of preterm brth in New York City, 2013–2017. Am J Public Health 2020;110(7):1046-53.

[35] Molitor J, Su JG, Molitor NT, Rubio VG, Richardson S, Hastie D, et al. Identifying vulnerable populations through an examination of the association between multipollutant profiles and poverty. Environ Sci Technol 2011;45(18):7754-60.

[36] Prochaska JD, Nolen AB, Kelley H, Sexton K, Linder SH, Sullivan J. Social determinants of health in environmental justice communities: Examining cumulative risk in terms of environmental exposures and social determinants of health. Hum Ecol Risk Assess 2014;20(4):980-94.

Section Two: Acknowledgments

Chapter 8

About the Editors

I Leslie Rubin, MD, is originally from South Africa, where he trained in pediatrics and emigrated to the United States in 1976. He is currently an Associate Professor in the Department of Pediatrics at Morehouse School of Medicine, Adjunct Associate Professor, Department of Pediatrics, Emory University School of Medicine, Director of Break the Cycle Program at the Southeast Pediatric Environmental Health Unit at Emory University (PEHSU), President and Founder of Break the Cycle of Health Disparities Inc, and Medical Director of The Rubin Center for Autism and Developmental Pediatrics in Atlanta, Georgia. He has been involved in the health care of children and adults with intellectual and developmental disabilities since 1977. In 1989, he and Allen Crocker published the first textbook on medical care for children and adults with developmental disabilities with the second edition in 2006. Sadly, Allen Crocker passed away in 2011, but, in partnership with Joav Merrick and others published the 3rd edition: "Health care for people with intellectual and developmental disabilities across the lifespan" with Springer in 2016. He currently directs interdisciplinary clinical programs for children with autism, cerebral palsy and other developmental disabilities and consults to a variety of programs including the Emory University Neurodevelopmental Exposure Clinic (ENEC), an interdisciplinary, research-based program looking at the impact of prenatal exposures on child development and behavior in the context of the social and economic disadvantage and health disparities. Among his awards for various activities, he received the Robert E Cooke Lifetime Achievement Award from the AADMD in 2015 and the Autism Achievement Award in 2016 from the Annual Conference and Exposition of Georgia, and the F Edwards Rushton CATCH Award from the American Academy of Pediatric in 2018, and most recently, was named one of 50 Advocate Heroes over the 50 years of Exceptional Parent Magazine. He is actively involved in the American Academy of Pediatrics as the Chair of the Georgia Chapter Environmental Health Committee and District X CATCH Facilitator for the AAP. He was on the US Environmental Protection Agency Board of Scientific Counsellors focusing on Sustainable Healthy Communities from 2016-2022 and is now on the National Advisory Council for Children and Disasters. He is also the Chair of the National PEHSU Steering Committee and is on the Advisory Councils of HERCULES, the Human Exposome Research Center at Emory University; and on the National START Advisory Committee, based at the University of New Hampshire. In 2004 he founded the Break the Cycle program which addresses children's environmental health disparities by recruiting bright university students to develop creative projects that will Break the Cycle and then publish their papers in international journals and books on public health. To date there have been more than 150 students from the USA, Latin America and Africa. The program received the Children's Environmental Health Excellence Award from the EPA in 2016. Email: lrubi01@emory.edu.

Joav Merrick, MD, MMedSci, DMSc, born and educated in Denmark is a professor of pediatrics, Division of Pediatrics, Hadassah Hebrew University Medical Center, Mt Scopus Campus, Jerusalem, Israel and Kentucky Children's Hospital, University of Kentucky, Lexington, Kentucky United States and professor of public health at the Center for Healthy Development, School of Public Health, Georgia State University, Atlanta, United States, the former medical director of the Health Services, Division for Intellectual and Developmental Disabilities, Ministry of Social Affairs and Social Services, Jerusalem, the founder and director of the National Institute of Child Health and Human Development in Israel. Numerous publications in the field of pediatrics, child health and human development, rehabilitation, intellectual disability, disability, health, welfare, abuse, advocacy, quality of life and prevention. Received the Peter Sabroe Child Award for outstanding work on behalf of Danish Children in 1985 and the International LEGO-Prize ("The Children's Nobel Prize") for an extraordinary contribution towards improvement in child welfare and well-being in 1987. In 2017 appointed a Kentucky Colonel by the Commonwealth of Kentucky, the highest honor the governor can bestow to a person. Email: jmerrick@zahav.net.il.

Chapter 9

About the Break the Cycle of Health Disparities Inc.

Break the Cycle of Health Disparities Inc. (BCHD) is a private, not-for-profit 501c3, non-governmental organization dedicated to exploring the social, economic and environmental determinants of health in the context of children's health disparities and to use this to promote health equity for all children.

BCHD collaborates with the Southeast PEHSU to produce the annual Break the Cycle of Children's Environmental Health Disparities Program (*Break the Cycle*). *Break the Cycle* is a multidisciplinary academic research and training program that focuses on raising awareness of children's environmental health disparities and cultivating future leaders among university students in the fields of medicine, nursing, public health, law, social work, education and other related disciplines.

Each year, *Break the Cycle* invites students from a diversity of colleges and universities to develop projects that will "break the cycle of children's environmental health disparities". Break the Cycle faculty guides and monitors the progress of the students towards compeltion of their projects. The students present their completed projects at a national conference and their work is published in international journals. Over the past 15 years Break the Cycle has engaged over 150 students from 50 different university departments in 12 states in the USA, as well as students from Latin America, Europe and Africa, that have resulted in the publication of 13 journal supplements. These sets of projects form the basis for this book series.

In 2016, *Break the Cycle* received a Children's Environmental Health Excellence Award from the US Environmental Protection Agency for the achievements of the program.

BCHD has expanded its areas of interest and practice to Break the Cycle of Autism Disparities, Break the Cycle of Climate Change for Vulnerable Children, Break the Cycle of Environmental Degradation, Break the Cycle of Children's Mental Health Disparities, and Break the Cycle of Health Disparities for Indigenous Children.

Contact

I Leslie Rubin MD
Associate Professor, Department of Pediatrics,
Morehouse School of Medicine
Adjuct Associate Profesor, Department of Pediatrics,
Emory University School of Medicine
Director, Break the Cycle Program, Southeast Pediatric Environmental Health Specialty Unit, Emory University School of Medicine
President and Founder, Break the Cycle of Health Disparities Inc
Medical Director, The Rubin Center for Autism and Developmental Pediatrics
750 Hammond Drive, Building 1, Suite 100, Atlanta, GA 30328,
United States of America
Email: lrubi01@emory.edu
Website: www.breakthecycleprogram.org
Website: www.therubincenterforautism.org

Chapter 10

About the Pediatric Environmental Health Specialty Unit (PEHSU)

The Pediatric Environmental Health Specialty Units (PEHSU's) are an interconnected system of regional centers located throughout the United States that respond to questions from public health professionals, clinicians, policy makers, and the public about environmental health concerns related to children and reproductive age adults. After twenty years of programmatic operations, collaborations, and a growing desire to protect the health of children and reproductive age adults from environmental hazards, the PEHSU Program has gained the attention and respect of the pediatric and reproductive environmental medical community and families at risk from exposure to environmental contaminants. The PEHSU programs have a critical role in supporting the mission of the Agency for Toxic Substances and Disease Registry (ATSDR) to prevent or mitigate the adverse human health effects and diminished quality of life that result from exposure to hazardous substances in the environment and the mission of the US Environmental Protection Agency (EPA) to protect human health and the environment. The PEHSUs provide a credible, academic, non-commercial source of health information and clinical expertise regarding environmental toxins and their effects on children and reproductive age adults.

ATSDR and EPA established the PEHSUs to increase the capacity of health care providers to identify and treat environmental-related health conditions, prevent exposure to children and reproductive age adults, and communicate health risks from environmental exposures. The PEHSU Network includes teams of medical specialists (pediatricians, medical toxicologists, obstetricians, nurses, nurse practitioners, occupational and environmental health physicians), public health professionals, researchers, psychologists, and educators to provide consultations, trainings, referrals and clinical services to health care providers, governmental agencies and the general public on pediatric and reproductive environmental health. PEHSUs also engage communities and stakeholders in public outreach projects.

The National PEHSU Network helps individuals and communities to make informed decisions regarding environmental health concerns. Common concerns include lead poisoning, fungus or mold, pesticides, phthalates, air quality and school safety. There have also been national responses generated for hazards from wildfires, hurricanes, oil spills, carbon monoxide, mercury spills, de-icer leaks at airports, and industrial pollution.

While Poison Control Centers respond to acute toxic exposures, the PEHSU Network has expertise in long term, chronic low dose exposures, group exposures, and the development and dissemination of public messages. The PEHSU Network partners closely with local, state, and federal agencies and non-profit professional organizations in carrying out this work.

The PEHSUs collaborate with ATSDR and EPA to develop fact sheets targeting clinicians and the public on issues associated with flooding, hurricanes, wildfires, oil spills, formaldehyde, carbon monoxide poisoning, arsenic in foods, medical management of lead exposure, phthalates and bisphenol A, melamine, and other issues. They edit academic textbooks, publish manuscripts, and conduct outreach and educational programs to enhance pediatric and reproductive environmental health literacy.

The PEHSU's provide online information (website: www.pehsu.net) with a variety of materials to educate professionals and the general public in reproductive and children's environmental health through conferences, presentations and social media. PEHSU members serve on national boards and committees to advance collaborative activities and community expertise in environmental health. The PEHSUs train and mentor medical, toxicology, nursing, and other graduate students and provide hands-on opportunities in pediatric environmental medicine and reproductive epidemiology. The PEHSUs have advised first responders, parents, health officials, and school administrators on how to assess and mitigate environmental emergencies and address health-related findings. They have also provided training to establish PEHSUs in Chile, Argentina, Canada, Mexico, Vietnam, and Israel.

The PEHSUs serve to translate research findings into clinical and public health practice. PEHSU faculty are prepared to recognize, treat, and prevent toxic environmental exposures in children and communicate risk, treatment, and prevention strategies in a meaningful and culturally appropriate way.

Southeast PEHSU

The Southeast PEHSU, within the National PEHSU network, serves the southeastern United States and includes Alabama, North Carolina, South Carolina, Florida, Georgia, Kentucky, Mississippi and Tennessee. The Southeast PEHSU is based in the Nell Hodgson Woodruff School of Nursing at Emory University in Atlanta, GA and works closely with Region 4 of the US Environmental Protection Agency (EPA) and the Agency for Toxic Substances and Disease Registry (ATSDR) at the Centers for Disease Control and Prevention (CDC). The team includes Abby Mutic, PhD, MSN, CNM (Assistant Professor and Certified Nurse Midwife), Leslie Rubin, MD (Developmental Pediatrician), Melissa Gittinger, DO (Medical Toxicologist), Rebecca Philipsborn, MD, MPA (Pediatrician), Victoria Green, MD, JD, MBA (Obstetrician Gynecologist), Henry Falk, MD MPH (Formerly Director of ATSDR), Nathan Mutic, MS, MAT, MEd (Educator and Program Administrator), and Wayne Garfinkel, BSCE (formerly of Region 4 EPA). In addition, Benjamin Gitterman, MD (Pediatrician) works with the Southeast PEHSU on the *Break the Cycle* program.

Contact Southeast PEHSU

Abby Mutic, PhD, MSN, CNM
Director, Southeast Pediatric Environmental Health Specialty Unit
Nell Hodgson Woodruff School of Nursing, Emory University
1520 Clifton Road, Atlanta, GA 30322, United States.
Toll free tel: (877) 33 PEHSU or (877) 337-3478
Tel: (404) 727-7980
E-mail: sepehsu@emory.edu
Website:
https://www.nursing.emory.edu/pages/southeastern-pediatric-environmental-health-specialty-unit.

Chapter 11

About the National Institute of Child Health and Human Development in Israel

The National Institute of Child Health and Human Development (NICHD) in Israel was established in 1998 as a virtual institute under the auspices of the Medical Director, Ministry of Social Affairs and Social Services in order to function as the research arm for the Office of the Medical Director. In 1998 the National Council for Child Health and Pediatrics, Ministry of Health and in 1999 the Director General and Deputy Director General of the Ministry of Health endorsed the establishment of the NICHD.

Mission

The mission of the National Institute for Child Health and Human Development in Israel is to provide an academic focal point for the scholarly interdisciplinary study of child life, health, public health, welfare, disability, rehabilitation, intellectual disability and related aspects of human development. This mission includes research, teaching, clinical work, information and public service activities in the field of child health and human development.

Service and Academic Activities

Over the years many activities became focused in the south of Israel due to collaboration with various professionals at the Faculty of Health Sciences (FOHS) at the Ben Gurion University of the Negev (BGU). Since 2000 an affiliation with the Zusman Child Development Center at the Pediatric Division of Soroka University Medical Center has resulted in collaboration around the establishment of the Down Syndrome Clinic at that center. In 2002 a full course on "Disability" was established at the Recanati School for Allied Professions in the Community, FOHS, BGU and in 2005 collaboration was started with the Primary Care Unit of the faculty and disability became part of the Master of Public Health Course on "Children and society". In the academic year 2005-2006 a one semester course on "Aging with disability" was started as part of the Master of Science program in gerontology in our collaboration with the Center for Multidisciplinary Research in Aging. In 2010 collaborations with the Division of Pediatrics, Hadassah Hebrew University Medical Center, Jerusalem, Israel around the National Down Syndrome Center and teaching students and residents about intellectual and developmental disabilities as part of their training at this campus.

Research Activities

The affiliated staff have over the years published work from projects and research activities in this national and international collaboration. In the year 2000 the International Journal of Adolescent Medicine and Health and in 2005 the International Journal on Disability and Human Development of De Gruyter Publishing House (Berlin and New York) were affiliated with the National Institute of Child Health and Human Development. From 2008 also the International Journal of Child Health and Human Development (Nova Science, New York), the International Journal of Child and Adolescent Health (Nova Science) and the Journal of Pain Management (Nova Science) affiliated and from 2009 the International Public Health Journal (Nova Science) and Journal of Alternative Medicine Research (Nova Science). All peer-reviewed international journals.

National Collaborations

Nationally the NICHD works in collaboration with the Faculty of Health Sciences, Ben Gurion University of the Negev; Department of Physical Therapy, Sackler School of Medicine, Tel Aviv University; Autism Center, Assaf HaRofeh Medical Center; National Rett and PKU Centers at Chaim Sheba Medical Center, Tel HaShomer; Department of Physiotherapy, Haifa University; Department of Education, Bar Ilan University, Ramat Gan, Faculty of Social Sciences and Health Sciences; College of Judea and Samaria in Ariel and in 2011 affiliation with Center for Pediatric Chronic Diseases and

National Center for Down Syndrome, Department of Pediatrics, Hadassah Hebrew University Medical Center, Mount Scopus Campus, Jerusalem.

International Collaborations

Internationally with the Department of Disability and Human Development, College of Applied Health Sciences, University of Illinois at Chicago; Strong Center for Developmental Disabilities, Golisano Children's Hospital at Strong, University of Rochester School of Medicine and Dentistry, New York; Centre on Intellectual Disabilities, University of Albany, New York; Centre for Chronic Disease Prevention and Control, Health Canada, Ottawa; Chandler Medical Center and Children's Hospital, Kentucky Children's Hospital, Section of Adolescent Medicine, University of Kentucky, Lexington; Chronic Disease Prevention and Control Research Center, Baylor College of Medicine, Houston, Texas; Division of Neuroscience, Department of Psychiatry, Columbia University, New York; Institute for the Study of Disadvantage and Disability, Atlanta; Center for Autism and Related Disorders, Department Psychiatry, Children's Hospital Boston, Boston; Department of Pediatric and Adolescent Medicine, Western Michigan University Homer Stryker MD School of Medicine, Kalamazoo, Michigan, United States; Department of Paediatrics, Child Health and Adolescent Medicine, Children's Hospital at Westmead, Westmead, Australia; International Centre for the Study of Occupational and Mental Health, Düsseldorf, Germany; Centre for Advanced Studies in Nursing, Department of General Practice and Primary Care, University of Aberdeen, Aberdeen, United Kingdom; Quality of Life Research Center, Copenhagen, Denmark; Nordic School of Public Health, Gottenburg, Sweden, Scandinavian Institute of Quality of Working Life, Oslo, Norway; The Department of Applied Social Sciences (APSS) of The Hong Kong Polytechnic University Hong Kong.

Targets

Our focus is on research, international collaborations, clinical work, teaching and policy in health, disability and human development and to establish the NICHD as a permanent institute in Israel in order to conduct model research and provide policy aspects in the field of interest.

Contact

Professor Joav Merrick, MD, MMedSci, DMSc
Director, National Institute of Child Health and Human Development, Jerusalem, Israel. E-mail: jmerrick@zahav.net.i.

Section Three: Index

Index

L

M

P

R